WAR DEPARTMENT TECHNICAL MANUAL
TM 4-210

This manual supersedes TM 4–210, 1 March 1940

SEACOAST ARTILLERY WEAPONS

WAR DEPARTMENT • 15 OCTOBER 1944

RESTRICTED: *Dissemination of restricted matter.*—The information contained in restricted documents and the essential characteristics of restricted material may be given to any person known to be in the service of the United States and to persons of undoubted loyalty and discretion who are cooperating in Government work, but will not be communicated to the public or to the press except by authorized military public relations agencies. (See also par. 23*b*, AR 380-5, 15 Mar 1944.)

Army Field Printing Plant
The Coast Artillery School
Fort Monroe, Va.

WAR DEPARTMENT,
WASHINGTON 25, D.C., 15 October 1944.

TM 4–210, Seacoast Artillery Weapons is published for the information and guidance of all concerned.

[A.G. 300.7 (1 Sep 44)]

BY ORDER OF THE SECRETARY OF WAR:

G. C. MARSHALL,
Chief of Staff.

OFFICIAL:
J. A. ULIO,
*Major General,
The Adjutant General.*

DISTRIBUTION:

SvC (10); Def Comds (3); IC & H 4 (3); Chiefs of Tech Sv (2); Arm & Sv Boards (2); Gen & Sp Sv Schs (10); Hq, PC & S having Seacoast Artillery Units (1).

IC & H 4: T/O 4-46; 4-47; 4-62; 4-66; 4-67; 4-68; 4-77; 4-152; 4-156; 4-157; 4-232; 4-240-1S; 4-260-1.

(For explanation of symbols see FM 21–6.)

CDSG Press
1700 Oak Lane
McLean, VA 22101 USA

The CDSG Press is part of the Coast Defense Study Group, Inc. (CDSG). The CDSG is a tax-exempt corporation dedicated to the study of seacoast fortifications. The purpose of the CDSG include educational research and documentation, preservation of historic sites and artifacts, site interpretation, and assistance to other organization in the preservation and interpretation of coast defense sites. Membership is open to any person or organization interested in the study of coast defenses and fortifications. Membership in the CDSG will allow you to attend the annual conference, special tours, and receive the CDSG's quarterly journal and newsletter. Please see the CDSG Membership and CDSG Press pages at the back of this reprint. Please visit the CDSG website at www.cdsg.org.

The CDSG Press would like to thank Terrance McGovern and Mark Berhow for their efforts in producing this reprint. Printed in the United States of America. For questions about this reprint, please contact Terry McGovern at tcmcgovern@att.net or at 1700 Oak Lane, McLean, VA 22101-3326 USA.

IBSN 978-0-9748167-4-6
(Paperback B&W)
LIBRARY OF CONGRESS CATALOG CARD NUMBER
20-18958732

Original Edition: 1944
Reprint Edition: 1995
2nd Reprint Edition: 2018

CONTENTS

	Paragraphs	Page
CHAPTER 1. General	1–3	1
CHAPTER 2. Cannon.		
SECTION I. Tubes	4–10	2
II. Breechblocks	11–16	16
III. Firing mechanisms	17–23	38
CHAPTER 3. Recoil and counterrecoil mechanisms	24–31	55
CHAPTER 4. Carriages.		
SECTION I. General	32	73
II. Fixed carriages	33–45	74
III. Motor-drawn carriages	46–48	135
IV. Railway artillery carriages	49–53	158
CHAPTER 5. Searchlights	54–56	172
CHAPTER 6. Demolition of materiel	57–64	184
APPENDIX I. Chart—Characteristics of seacoast artillery weapons		188
INDEX		189

This manual supersedes TM 4-210, 1 March 1940

CHAPTER 1

GENERAL

1. Purpose
The purpose of this manual is to acquaint the coast artilleryman with the principal types of weapons employed in seacoast artillery.

2. Mission
The mission of seacoast artillery is to protect harbors, naval installations, and the coast line against hostile attack by the employment of artillery fire and submarine mines. Seacoast artillery embraces all fixed and mobile artillery assigned to harbor and coastal defense.

3. Scope
  a. This manual deals with the basic principles employed in the construction and operation of all types of seacoast artillery weapons. For study and simplicity of presentation, weapons are discussed according to their principal parts. Searchlights are treated briefly.
  b. The coast artilleryman should supplement his study of this general account of weapons and searchlights by consulting the more detailed publications on the particular armament to which he is assigned. (See FM 21-6.) No description of position finding and pointing equipment has been given, as this information can be found in FM 4-15. Antiaircraft artillery and submarine mines are not discussed.

NOTE. For definitions of military terms not contained in this manual, see TM 20-205.

CHAPTER 2

CANNON

Section I. TUBES

4. History

a. Mechanical devices for the hurling of missiles were conceived early in the history of warfare. Dating as far back as the *Old Testament,* mention is made of the use of massive "engines of war" designed to throw huge stones to breach enemy fortifications. The use of these devices, such as the ballista (fig. 1) and the catapult, was gradually discontinued with the innovation of the cannon. The first date of employment of the cannon cannot be accurately ascertained.

Figure 1. Ballista.

b. The modern cannon has had a colorful evolution. Prior to about 1860 it was usually cast in one piece. As the weight of powder charges increased, so too did the weight of the cannon. This additional weight impeded the cannon's mobility and tactical employment. Furthermore, it was found by trial and error that an increase in thickness of the cannon wall did not give a corresponding increase in strength. Examination of cannon developing enlarged bores due to overstress disclosed that the part of the wall next to the bore always broke down first. This indicated that the inner part of the wall was failing before the outer part and that the outer part was not absorbing its share of the stress (par. 8) when the cannon was fired. To overcome these difficulties, a cannon capable of withstanding tangential or circumferential stress (par. 8a) had to be constructed in such a manner that the layer of metal near the bore would be under compression by the outer layers when the cannon was at rest. An American, General T. J. Rodman, contributed to the strength of the cannon by casting it upon a hollow core while cooling the inner surface with a stream of water. As a result of this process, the interior metal about the bore of the Rodman gun (fig. 2)

Figure 2. The 15-inch Rodman gun.

was in a state of compression and the exterior in a state of tension. Another American, Robert P. Parrott, attained the same result by shrinking on a jacket. From these developments has come the modern built-up cannon consisting of two or more concentric cylinders assembled by the shrinkage methods. Two other important developments were the invention of a safe breech-loading cannon, which decreased loading time, and the introduction of rifling, which imparted a gyroscopic spin to the projectile with resultant increase in range and accuracy. Some of the earliest cannon were breech loaders, but they were discarded in favor of the muzzle loaders, which remained in use for over three hundred years until the modern breechblock, with an effective means of preventing the escape of gases to the rear, was devised. With the modern breech-loading cannon, came more effective rifling.

c. In the European War, 1914-1918, the Germans at first secured a tremendous advantage through their successful employment of the large howitzer as a field and siege weapon. It possessed high-angle fire, good range at medium velocity, maximum life, and considerable mobility. To cope with this situation, the Allies introduced the large scale use of tractor-drawn and railway artillery. Longer ranges and extensive mobility were of paramount importance to both sides to enable them to cripple hostile supply and communication centers far behind the lines.

d. Since 1918, the great advances in cannon construction have led to greater mobility, increased range, and rapidity of fire. The improvements are well exemplified in our modern 155-mm and centrifugally cast 90-mm guns.

5. Types

a. Cannon are classified into three types: gun (rifle), howitzer, and mortar (fig. 3).

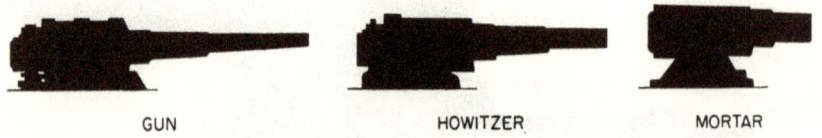

GUN   HOWITZER   MORTAR

Figure 3. Silhouettes of a gun, howitzer, and mortar.

(1) *Gun (rifle).* The gun (fig. 32) possesses a relatively long barrel, low angle of fire, and high muzzle velocity. Unlike the mortar, it uses relatively few different powder charges. Some modern guns have a maximum elevation of 65° while others, such as the 8-inch gun Mk. VI Mod. 3A2 on barbette carriage M1 (max. elevation 45°), have less. The minimum elevation is approximately 0°, but some guns, such as the 90-mm gun, may be fired at elevations below 0°. The great length of the gun allows the powder gases to act for a relatively long period on the projectile, thus producing a high muzzle velocity. The modern gun has several of the advantages of the howitzer without its disadvantages. In an effort to decrease erosion, which is the gradual wearing away of the surface of the bore, powder charges of different weights are furnished for some guns. Also, the different weights give the commander of a battery of guns a choice of most of the alternatives formerly possessed by the howitzer and mortar commander. To hit a certain point, he may use a heavy powder charge and a low angle of elevation, a medium powder charge and a medium angle of elevation, or a small powder charge and a high angle of elevation. Within the ranges covered by all powder charges, a battery commander shelling

ships that are particularly vulnerable to high-angle fire may use a small powder charge. If firing against ships deficient in side armor, he may use the heavy charge. (See FM 4-5.)

, (2) *Howitzer*. The howitzer (fig. 94), used more extensively by the field artillery, is a cannon of relatively large caliber (see *b* following) in proportion to its medium length. The firing elevations range between 0° and 65°. In comparison to the mortar, it uses a relatively small number of different powder charges. Few howitzers are found in modern seacoast defense.

(3) *Mortar*. The mortar, a relatively short weapon, develops a low muzzle velocity and is designed to fire at high elevations between 45° and 65°. The seacoast mortar has proved to be ineffective in the present world conflict because of its vulnerability, comparatively short range, and long time of flight of its projectile. Also, the mortar is a zone weapon which employs several different weights of powder charges and two weights of projectiles. Because of this, it presents great difficulties in directing fire on modern fast-moving targets passing rapidly from one zone to another. It is now considered obsolete.

b. Cannon are also classified as to caliber. The caliber of a cannon is the diameter of its bore between opposite lands (fig. 5). Major caliber weapons, or primary armament, are 12 inches or more in caliber, while those less than 12 inches are classified as minor caliber or secondary armament. Caliber is also used to express the over-all length of cannon; that is, it expresses the length of the tube containing the powder chamber and rifled bore. For example, "a 14-inch, 50-caliber gun" is one which has a bore with a diameter of 14 inches and a length of 50 calibers (50 $\times$ 14 inches or 700 inches).

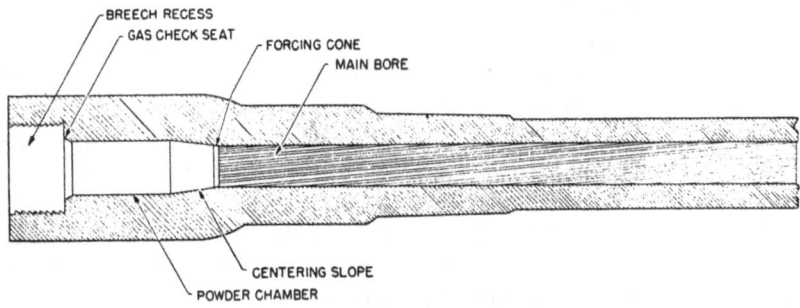

Figure 4. Subdivisions of the interior of a cannon.

6. Subdivisions of the Interior

For the sake of simplicity, this discussion divides the interior of the modern cannon into six parts (fig. 4).

a. The breech recess has the largest inside diameter of all the open sections and contains the threaded sectors in which the breechblock operates to close the rear end of the piece.

b. The gas check seat is that portion of the bore especially designed to fit the gas check pad and split rings. With them, it provides a gas-tight seal to prevent the escape of gases to the rear when firing (par. 12).

c. The powder chamber is made larger in diameter than the remainder of the bore to obtain greater powder volume in a shorter length.

d. The centering slope is the forward end of the powder chamber. As its name indicates, its purpose is to lift the projectile from the level of the powder chamber floor and center it in the main bore. This is accomplished by the gradual slope of the wall, formed by the contraction of the bore. It acts in much the same way as a funnel.

e. The forcing cone is made by cutting away the rifling at the rear of the main bore. Here the greatest wear takes place. Like the centering slope, it assists in centering the projectile in the bore. It also provides the surface on which the rotating band stops the forward motion of the projectile as it is rammed.

f. The main bore extends from the forcing cone to the muzzle and is rifled throughout. It gives rotation to the projectile as it is forced forward by the expansion of the gases generated by the burning propelling charge.

7. Rifling

a. Rifling, which consists of a number of helical grooves cut into the bore of the cannon, is designed to impart a gyroscopic effect or spin to the projectile as it is forced through the bore (fig. 5). This rotating motion permits the effective use of an elongated projectile by eliminating the end-over-end flight given it by smooth-bore cannon. Rifling, by insuring a steady "point first" flight of the projectile, is responsible for minimum air resistance, greater uniformity of flight, and greater accuracy. The number of revolutions per minute necessary to accomplish this result is a definite value which depends on the projectile and the muzzle velocity used; the higher the muzzle velocity and the longer the shell (with reference to its diameter) the greater the r.p.m. required. On certain small, high-velocity weapons, speeds approaching 20,000 r.p.m. are necessary. For a 16-inch gun, the velocity of rotation is about 4,050 r.p.m.

b. The accuracy life of a weapon depends primarily on the condition of the rifling. The number, width, and profile of the lands (the surfaces of the bore between the grooves) must be such as to withstand the stresses set up as the projectile passes through the bore.

Figure 5 illustrates the rifling now standard in modern cannon. Several different profiles of lands and grooves have been tried, but at present the rib rifling with equal widths of lands and grooves is generally used.

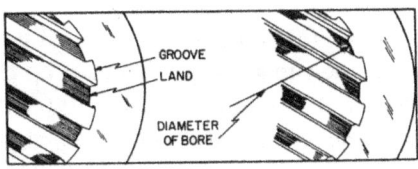

Figure 5. Modern rib rifling.

c. Twist or steepness of rifling at any point is the inclination of the grooves with respect to the axis of the bore. It may be uniform, increasing, or a combination of the two. Uniform twist is now specified for practically all Army cannon. Rifling steepness and direction are usually designated, for example, as "right-hand, uniform, one turn in 25 calibers" or "one turn in 40 calibers." Thus, for a 12-inch gun, one turn in 25 calibers means one turn in 25 feet of gun length (25×12 inches equals 300 inches or 25 feet).

d. It is necessary at this point to mention the part the rotating bands of the projectile play in conjunction with the rifling in the bore. As the projectile enters the main bore, the lands bite into the soft copper bands on the projectile and give the necessary rotation to the projectile while passing through the bore. Because of the close meshing of the bands into the lands and grooves, they also act as a gas seal, preventing the escape of gas from the powder chamber past the projectile. This sealing prevents undue erosion and eliminates loss of muzzle velocity which would occur if such an escape of gases took place. In addition, they serve as a means of holding the rammed projectile in the forcing cone regardless of the elevation to which the cannon is raised after loading.

8. Manufacture

a. GENERAL. (1) *Stress.* When a cannon is fired, the powder pressure subjects it to two stresses. One is a circumferential or tangential stress or tension coupled with a radial stress (fig. 6), tending to split the gun open longitudinally. The other is a longitudinal stress which tends to elongate the gun, but this is relatively unimportant compared to the first. In connection with cannon construction, artillerymen should keep in mind the following:

(*a*) A cannon is basically a tube capable of discharging a projectile at a high velocity. The tube is designed to withstand a given pressure from within.

(b) The stress to which a cannon is subjected should not exceed its elastic limit (the greatest unit stress the cannon is capable of withstanding without permanent deformation).

(c) When this elastic limit is exceeded, the cannon will be subjected to strain which is permanent deformation or change in dimension and shape.

(d) Experience (par. 4b) has shown that the use of increased powder pressure in a one-piece cannon cannot be met by increasing the thickness of the wall, because failure (strain) occurs in the metal near the bore before the outer metal receives its share of the stress.

(e) The problem is to make the outer layers take a proper proportion of the stress.

(f) Modern cannon are constructed in such a manner that the metal near the bore is under compression when the cannon is at rest.

(g) Allowable powder pressure may be increased in modern cannon because part of the pressure acts to relieve this initial compression and additional pressure is required to take the metal to its elastic limit.

(h) Since the initial compression of the metal near the bore is caused by the tension of the outer metal, the outer layers receive their share of the stress when the cannon is fired.

(i) Construction as well as size determines the strength of a modern cannon.

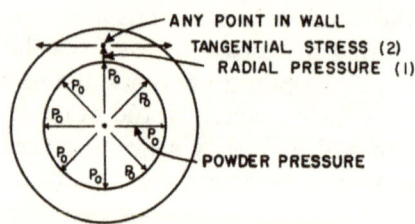

Figure 6. Diagram showing stresses in cannon walls.

(2) *The ideal cannon.* If it were practical to manufacture the theoretically ideal cannon, it would be made of an infinite number of very thin cylinders, each shrunk over the next smaller cylinder, and each in turn exerting a compression on the smaller or inner cylinders. While such a cannon was at rest, an initial state of compression would exist which would have to be overcome in firing before stress in a tangential direction reached the elastic limit of the wall. Therefore, the amount of powder pressure could be considerably increased, since part of this pressure would act against initial compression. In the theoretically ideal cannon, all layers of

the gun would be equally stressed and each cylinder would carry its share of the burden when the cannon was fired. Such a cannon wall would be stronger than a solid wall of the same thickness.

(3) *Manufacturing processes.* Since it is impracticable to produce such an ideal cannon, efforts have been made to reproduce these conditions as closely as possible. The effect can be partially achieved by shrinking a series of thin cylinders or hoops over each other. In order of their development, the four processes of modern cannon manufacture are as follows:

(a) Built-up process.
(b) Wire-wrapping process.
(c) Cold-working (autofrettage or radially expanding) process.
(d) Centrifugal-casting process (plus cold working afterward).

b. BUILT-UP PROCESS. (1) A built-up cannon consists of two or more concentric cylinders in which the inner tube is reinforced by means of jackets or hoops. The exterior parts are first heated until they will fit over the interior part and then are shrunk in position by the cooling process. The interior part is kept cool by water forced through it; the exterior parts are cooled from the breech end forward by spraying with water to insure proper seating on the inner cylinder.

(2) In the assembly of the 16-inch howitzer (fig. 7), the outer or B-tube is machined to its finished dimensions. A jacket and the A-hoop, whose inner diameters are slightly less than the outer diameter of the B-tube, are expanded by heating and fitted over the B-tube. Then, as they are cooled, the jacket and A-hoop shrink into place and compress the B-tube. In a similar manner, the B-hoop is shrunk over all these cylinders, giving more compression to the powder chamber of the cannon and adding weight to bring the center of gravity to the rear. The cannon is now complete except for the inner or A-tube. To insert this, the assemblage is heated and the cold A-tube is lowered into place.

(3) Normally, the tube containing the rifling wears out sooner than the rest of the cannon. When the rifling has become so worn as to interfere seriously with the accuracy of fire, the cannon can be dismounted and returned to the arsenal for retubing. For some small caliber guns, especially antiaircraft guns, a removable liner has been developed which can be installed by troops in the field. However, the centrifugal-casting process of manufacture (see *e* following) has made it possible to build a complete gun of small caliber almost as cheaply as to build a gun with a removable liner. Such centrifugally cast, monotube guns are replaceable by personnel in the field.

c. WIRE-WRAPPING PROCESS. (1) It is unlikely that any more wire-wound cannon will be built by the Army. However, since there are

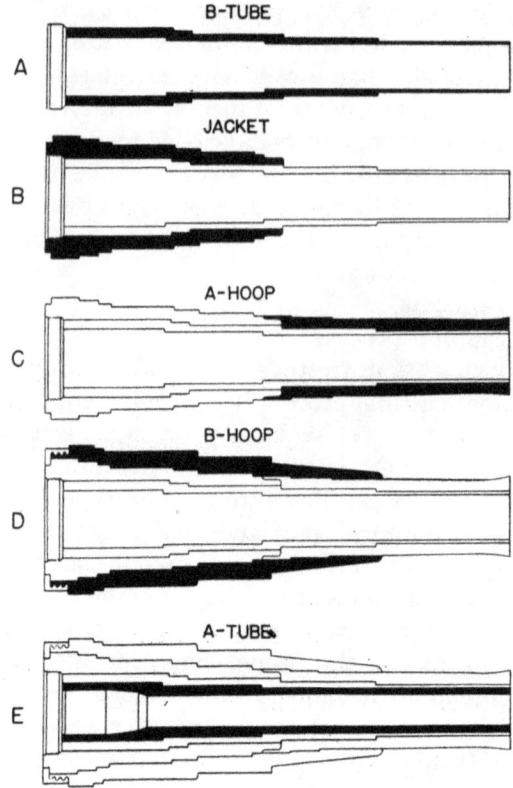

Figure 7. Order of assembly, built-up cannon, 16-inch howitzer M1920.

guns of this type in the service, a brief description of the wire-wrapping process will be given (fig. 8). Work starts at the B-tube, about which may or may not be shrunk a jacket. Over the B-tube a machine winds wire, square in cross section and about 1/10 inch wide, of great tensile strength. The wire places the tube under compression, producing the same result as the shrinking process used in built-up cannon assembly. More layers of wire are wound over the powder chamber than over the rest of the bore as this section demands greater initial compression. Hoops are shrunk over the layers of wire. The A-tube containing the rifling is then inserted inside the whole as in the built-up process.

(2) While the wire-wound gun represents the lightest major caliber gun today, it has less longitudinal rigidity and consequently more droop than other types. In addition, the gun is much more difficult to reline.

d. Cold-working Process. (1) In the cold-working or autofrettage process (fig. 9), a single cylinder, with an interior diameter

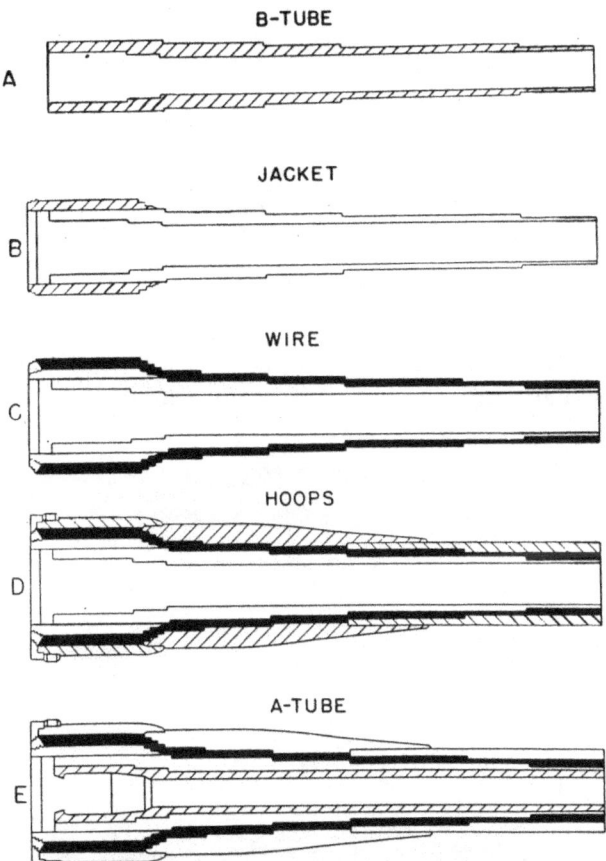

Figure 8. Order of assembly, wire-wrapped cannon, 16-inch gun M1919M2.

slightly less than desired, is subjected to hydraulic pressure (44,000 to more than 100,000 pounds per square inch) in a container shaped to the exterior of the gun. Autofrettage literally means "self-hooping." Permanent expansion is produced through overstressing the tube. However, this permanent set within the metal progressively diminishes as the outside of the tube is approached; that is, the fibers of the gun are strained in proportion to their distance from the bore. The outside diameter is enlarged very slightly as it expands against the sides of the container. After the removal of the pressure, the inner portions do not return to their original form. The outer layers remain more or less in a state of tension and act compressively on the inner layers, thus producing the same compressive effect as hoops. Since this process subjects the inner fibers of the gun to a stress beyond their elastic limits by pressure far in excess of any they will receive when the gun is fired, a piece is produced (with proper heat treatment) which will withstand a higher pressure than a built-up gun of the same caliber and weight. After this cold working, the gun is machined to its final dimensions and rifled as in any other process.

(2) Since its first development shortly before the European War, 1914–1918, the process has been improved to such an extent that manufacturing costs have been cut as much as 40 percent. As this process reduces the number of castings required, time and labor are saved. Another factor of paramount importance for mobile and rapid-fire guns is the reduction of weight accomplished by this method of manufacture. In this country the cold-working process has been applied to one-piece guns up to 6-inch or 155-mm caliber, and to 8-inch Navy guns of two-piece construction. The latter weigh 14 tons less than built-up guns of the same size and strength.

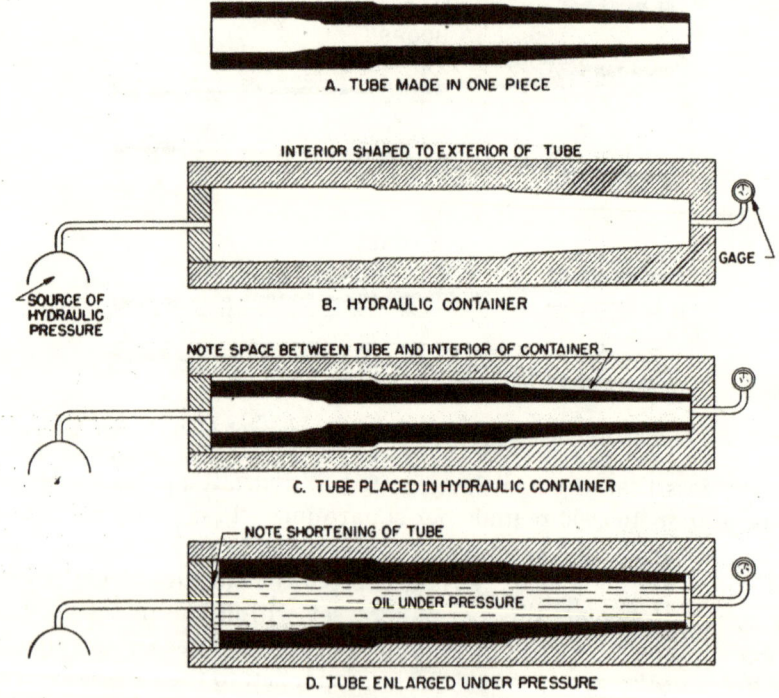

Figure 9. Cold-working process (container method).

e. CENTRIFUGAL-CASTING PROCESS. Recently, a process has been invented for casting guns centrifugally (fig. 10) in a machine similar to that used in making cast-iron pipe. While the cylindrical mold is rotated at high speed, molten steel is poured in at one end. Acting as any liquid would under centrifugal force, the molten metal hugs the surface of the mold and forms a hollow cylinder of molten metal.

During the cooling process, and while the mold is spinning, two important effects of centrifugal force on the molten metal take place. First, high outward pressure engendered in the metal by centrifugal force prevents, during cooling, the formation of blowholes and cracks, and forces impurities to the inner surface of the casting. Second, the same outward pressure produces a desirable graduation in the metallic structure of the casting by making it sufficiently hard inside (at the bore) to resist wear, and sufficiently ductile outside (next to the mold) to give great strength and elasticity. After cooling, the casting is ready for the slight machining required and for further heat treatment and cold-working as is necessary. Recent perfection of the process has enabled manufacturers to cut production time in half. This process has also been employed in the manufacture of our latest monotube 90-mm gun (fig. 32).

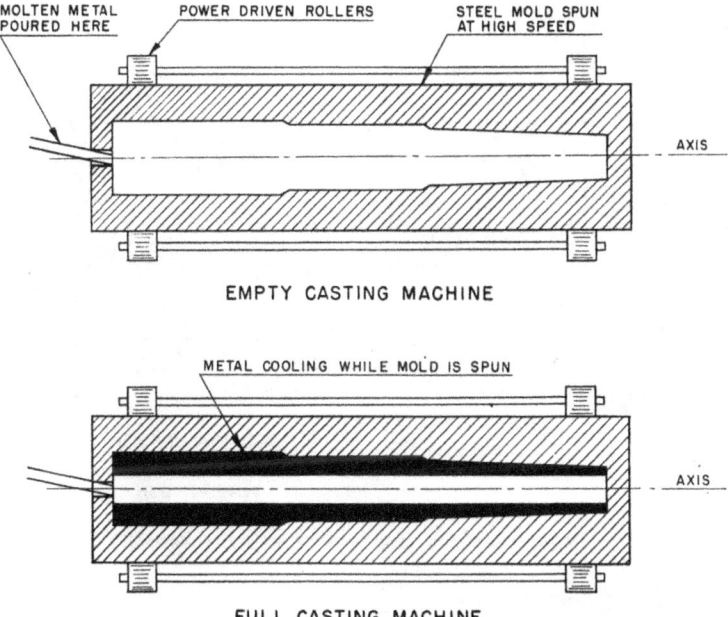

Figure 10. Centrifugal-casting process.

9. Support

Early in the development of artillery, it was found desirable to be able to alter the range of cannon by changing the elevation. Pivoting the cannon on trunnions proved to be the most suitable method of accomplishing this purpose. At first the trunnions were cast integrally with the cannon (fig. 2), but as recoiling mechanisms developed, this method was superseded by the use of long splines or slide rails

13

fastened to the cannon and parallel to the axis of the bore. These splines or slide rails (fig. 32) fit corresponding slots in a cradle (fig. 131). They allow the cannon to recoil and counterrecoil and prevent it from rotating to the left as a reaction of the rotation of the projectile to the right. On the outside of the cradle are mounted trunnions (fig. 83) which permit the cannon and cradle to be set in elevation, and recoil to take place parallel to the axis of the bore regardless of the elevation of the cannon. It is customary to have the trunnions located at or near the center of gravity of the cannon so that it will be "breech heavy" when loaded and "muzzle heavy" after firing. This balance makes it easier to elevate and depress the cannon. When the trunnions are placed well to the rear, as in the 90-mm and the 155-mm M1 guns (figs. 126 and 144), equilibrators are provided to counterbalance the unbalanced weight of the barrel assembly. Muzzle heaviness of the 8-inch gun Mk. VI Mod. 3A2 (fig. 97) is equalized by a counterweight.

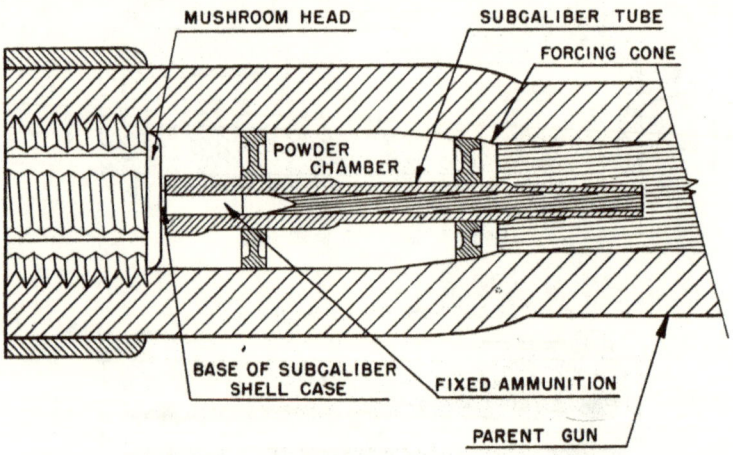

Figure 11. Subcaliber gun, interior-mounted, schematic drawing.

10. Subcaliber Guns

a. Because of the expense incurred in firing service ammunition, practice firing is often conducted with guns of smaller caliber. Some subcaliber guns are mounted in the bore (fig. 11) of the larger cannon; some are mounted on the outside (fig. 12). The type mounted inside the bore is merely the barrel of a gun with the necessary accessories for attaching it to the inside of the parent piece. As it has no breechblock or recoil system, the parent cannon absorbs the shock of firing. There is just enough clearance between the breech-

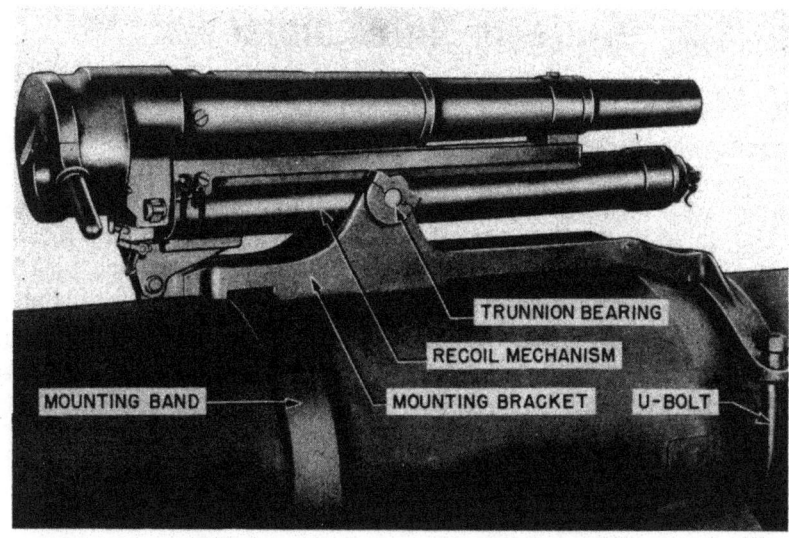

Figure 12. 37-mm subcaliber gun M1916, mounted on a 155-mm gun.

block of the parent cannon and the rear end of the subcaliber tube for the base of the cartridge case of the fixed ammunition. On the other hand, the exterior-mounted subcaliber gun has a complete recoil system and breechblock and is fired independently as is any small caliber gun using fixed ammunition (fig. 13).

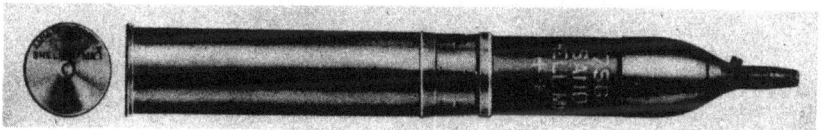

Figure 13. Subcaliber ammunition Mk. I, 75-mm gun, used on 16-inch gun.

b. There are several sizes of subcaliber guns, such as the caliber .30 tube for the 3-inch guns; 37-mm gun mounted on the 155-mm G.P.F.; and 75-mm gun mounted on the 16-inch guns and howitzers.

c. The Ordnance Department has under development 75-mm subcaliber guns for all modern seacoast fixed and mobile artillery of 6-inch and 155-mm caliber and above. These tubes are to be inserted in the chamber of the parent cannon. The ammunition to be used is the 75-mm projectile M48, inert loaded. Normal charge only will be provided. The standard firing circuit and firing mechanism of the parent cannon will be used when firing these subcaliber guns.

Section II. BREECHBLOCKS

11. General

Muzzle-loading methods remained in use long after the innovation of the breech-loading cannon because of the difficulties encountered in providing an effective and safe means of obturation. Not only is it necessary that the breech mechanism close the rear end of the cannon, but it is also of paramount importance that it prevent the escape of gases to the rear. Additional requirements of an efficient breech mechanism are: ease and rapidity of operation, safety, durability, ease of repair, and adaptability to mass production. These features are essential to facilitate the proper combat performance of modern weapons.

12. Obturation

a. GENERAL. Obturation is the prevention of the rearward passage of powder gases through the threads and other parts of the breech. These gases, which are developed by the rapid combustion of the powder in the powder chamber, have great velocities and exceedingly high temperatures. If allowed to escape to the rear, they would soon erode and ruin the breech mechanism. This escape of gases would also decrease the developed muzzle velocity of the cannon.

b. FIXED AMMUNITION. In guns that use fixed ammunition, such as the 90-mm and the 3-inch seacoast rapid-fire guns, obturation is performed by the expansion of the cartridge case under pressure of the burning powder to form a tight seal against the walls of the powder chamber. The breach of an ordinary rifle employs the same principle of obturation. With this brass-case system of obturation, a simple form of breechblock is often used, its only function being to hold the brass case in place against the pressure of the powder. Because of the weight and size of a round of ammunition, this system is not generally used in larger cannon. The total expansion of the shell case should not exceed the elastic limit of the metal, as otherwise the case may be permanently deformed or split and thus offer too much resistance to ejection.

c. DEBANGE SYSTEM. (1) Most countries use the DeBange system of obturation in larger cannon with separate loading ammunition. The main parts of the DeBange obturator are: a steel mushroom head and spindle; a gas check pad composed of an asbestos wire cloth cover filled with a plastic mixture of asbestos, tallow, and paraffin which melts only at very high temperatures; two split rings, of triangular cross section, ground to encircle the gas check pad; an inner ring for sealing around the spindle; and a filling-in disk. These parts are arranged as shown in figures 14, 15, and 31. When not under powder pressure, the split rings make light contact with the gas check seat of the cannon, allowing free closing of the breechblock.

(2) When the cannon is fired, the gases press against the mushroom head of the obturator spindle (fig. 14). This action compresses

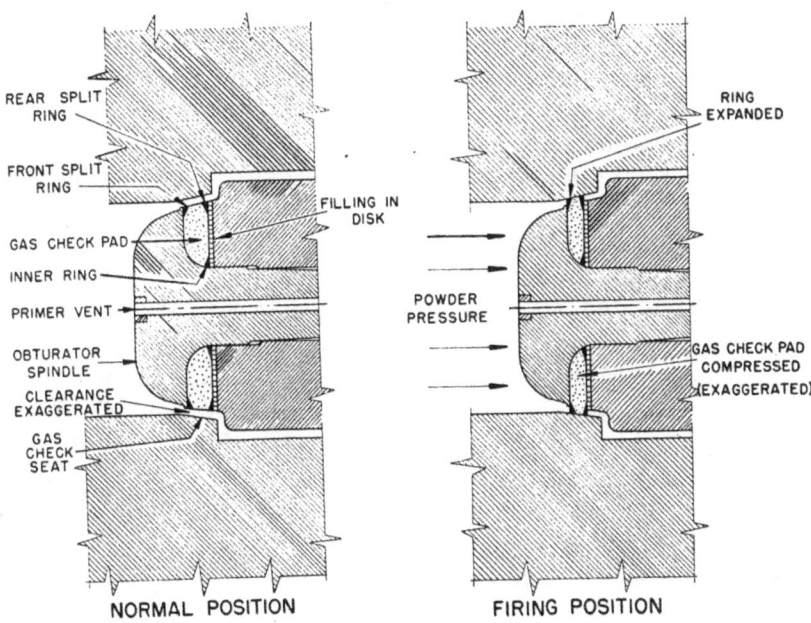

Figure 14. Operation of DeBange obturator, schematic drawing.

the gas check pad, causing it to expand radially. The expansion of the pad presses the split rings so that they increase in diameter and press firmly against the walls of the gas check seat to form a gas-tight seal. After firing, the pad is relieved of its pressure and returns

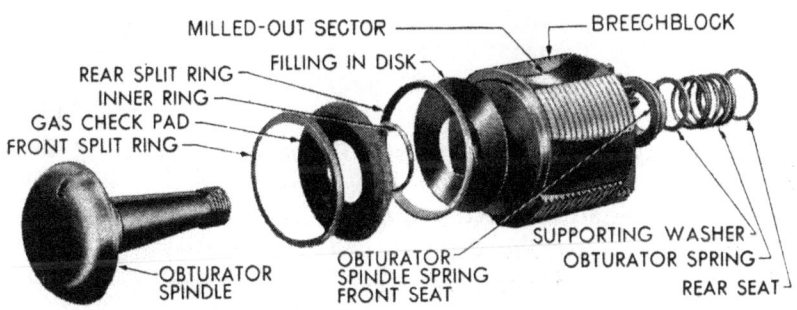

Figure 15. Obturator assembly, 155-mm gun G.P.F., exploded view.

to its original shape, the split rings contract to their normal diameter, and the breechblock is free to be opened for further firing. Sometimes, after several rounds, the gas check pad sticks to the gas check seat, making the opening of the breech difficult. However, the rotation of the block is not affected by the sticking because the gas check pad, split rings, and spindle remain stationary while the block rotates during closing and opening. The filling-in disk between the gas check pad and breechblock is kept well oiled to provide an oiled bearing surface for the rotation of the block. The small inner ring around the spindle prevents the escape of gas check pad composition at this point. The circular primer vent in the obturator spindle permits the passage of flame from the primer to the powder chamber to ignite the propelling charge when the cannon is fired.

13. Classes of Breechblocks

While there are many kinds of breechblocks, most of them can be classified under one of three general classes: (1) the slotted- or interrupted-screw, (2) the sliding-wedge, and (3) the eccentric-screw. The eccentric-screw class is too heavy for large cannon and is not generally used in seacoast artillery. The best example of this type is the breech mechanism of the famous French 75-mm gun M1897. The sliding-wedge is found in our 90-mm gun and is described in paragraph 15. It is readily adaptable to automatic features of weapons built for a high rate of fire. The slotted-screw class of breechblocks with the DeBange obturator system, described in paragraph 14, is the most common in our seacoast weapons.

14. Slotted-screw Breechblocks

a. GENERAL. The slotted- or interrupted-screw type of breechblock is used because it possesses several advantages: strength, rapidity of operation, reduction of weight in the breech section, adaptability to means of securing obturation with separate-loading ammunition, and uniform distribution of longitudinal stress produced by powder pressure developed in firing. However, it has not been effectively adapted to rapid automatic operation as has the sliding-wedge type used on our 90-mm gun.

b. METHODS OF THREADING. (1) The operation of a simple, plain-type, slotted-screw breechblock bears a resemblance to the engagement of a bolt and nut (fig. 16). If the threads of the nut are cut away from one-half of its circumference, and if the threads of the bolt are cut away in the opposite half of its circumference, then the bolt can be pushed through the nut without engaging any threads whatsoever. Locking may then be accomplished by rotating the bolt a simple half turn to engage the threaded sector of the bolt with the threaded sector of the nut. This arrangement allows the bolt to slide all the way home in one pushing movement, and it requires only a half turn to engage the bolt with the nut. The convenience of this arrangement is apparent if one considers the several turns that are required to lock an ordinary bolt and nut. The action of the bolt engaging the nut represents the action of the breechblock engaging the breech recess in a simple slotted- or interrupted-screw mechanism.

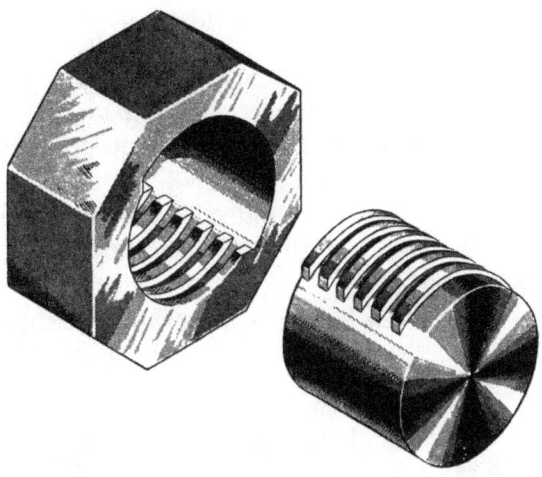

Figure 16. Principle of plain-type, slotted-screw breechblock, illustrated by nut and bolt.

(2) In the plain type of breechblock there is an equal number of slotted and threaded sectors (fig. 17). In this type only half of the circumference (180°) is threaded. In the Welin or step-threaded breechblock (fig. 29), there are three threaded sectors to one slotted sector. It may thus be seen that when the plain-type breechblock is completely closed, one-half of the circumference is meshed with the corresponding threads in the breech recess, whereas, under the same circumstances, with the three-step Welin type of breechblock, three-fourths of the circumference is meshed. This means that with the Welin system the block can be made smaller in diameter, shorter, and lighter than an old-type plain block of equivalent strength. In American cannon the step-threaded sectors are built in two, three, or four steps.

c. TYPES. Slotted-screw breechblocks may be classified according to their mechanisms into two types—the tray-supported (3-cycle) type and the carrier-supported (2-cycle) type.

(1) *Tray-supported.* (a) The tray-supported type of breech mechanism (figs. 27, 28, and 29) must accomplish three motions in

closing or opening the block. Starting from a closed position the mechanism must—

    1. *Rotate* the block until the threaded sectors are disengaged.
    2. *Translate* the block to the rear out of the breech recess onto a tray.
    3. *Swing* the block and tray from the rear of the breech recess, leaving the breech completely free for loading.

(b) When the block is locked in the closed position, there is no mechanical connection between the block and the tray. The sole purpose of the tray is to support the block when it is in the withdrawn position. The tray-supported type includes most of the older major-caliber blocks of conventional design. One of these is the breechblock, described in e following, used on the 12-inch gun. Another example is the breechblock used on the 8-inch gun Mk. VI Mod. 3A2 described in ƒ (4) following.

(2) *Carrier-supported.* (a) In the carrier-supported type (fig. 17) the opening or closing is generally accomplished in two cycles. The mechanism must—

    1. *Rotate* the block until the threads are disengaged.
    2. *Swing* the block out of the breech recess and clear of the breech.

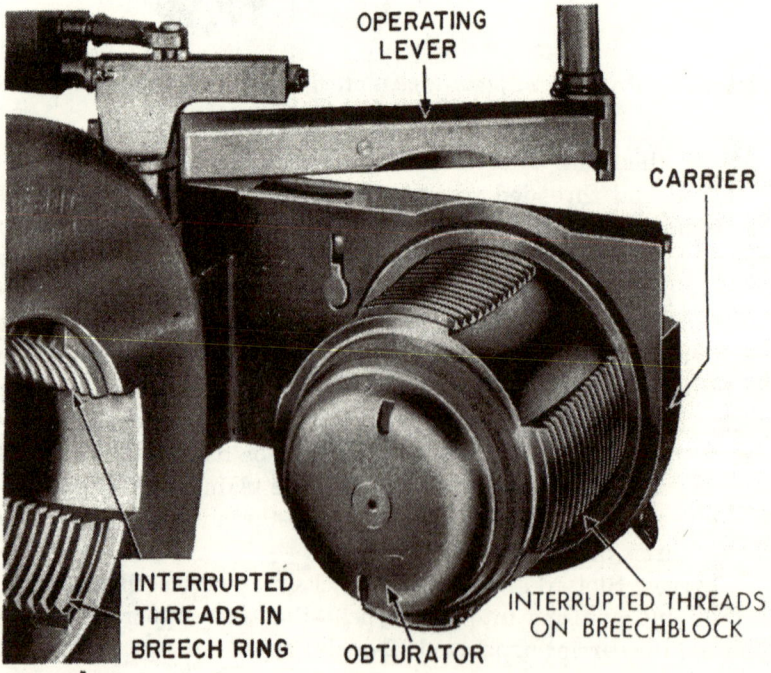

Figure 17. Breechblock showing interrupted threads, 155-mm gun G.P.F.

(*b*) In this type, the block is supported by a carrier with a hub (pintle). The hub, together with the obturator spindle, forms an assembly around which the block itself rotates, as a wheel around an axle. The hub also incloses the firing mechanism housing, obturator spindle, and its allied parts. Formerly this type of support was found only on small caliber rapid-fire guns, but modern design has incorporated this principle in blocks of the largest caliber. Blocks using this principle must be specially shaped so that they will clear the breech recess during the swing cycle. They must necessarily be short compared with their diameter, and some must have milled-out areas (fig. 15) in the slotted sectors of both the block and the breech recess to permit free movement into and out of the breech recess. The carrier-supported type contains breechblocks of many different designs such as the lever-pull mechanism of the 155-mm gun G.P.F. (see *d* following), the down-swing block used on the modern 16-inch gun (see *f* (3) following), and the tapered and ogival breechblocks found on some of the 6-inch guns (see *g* following).

d. BREECHBLOCKS, 155-MM GUN G.P.F. AND 6-INCH GUN M1900. (1) *Description.* The plain slotted-screw, lever-pull, and carrier-supported design is best exemplified by the breechblock on the 155-mm gun G.P.F. (fig. 17). Half the surface of this breechblock is plain and half is threaded; that is, it has four threaded sectors and four plain sectors. Similarly, the threads of the breech recess are cut away opposite the threaded sectors of the block. Because of this arrangement, the breechblock can be slid nearly to its home position with the threaded sectors of the block sliding in the blank sectors of the breech recess. With 1/8 of a turn, the block is rotated so its threads engage the threads of the breech recess, thus completely locking the block in place. Two slotted sectors are milled out to permit the block to clear the breech recess when the breechblock is swung in and out.

(2) *Operation.* The breechblock is screwed onto the front of the block carrier. The carrier swings to the right on a hinge pin. In this type of breech mechanism, the operation of rotating the block and swinging it clear of the breech recess is accomplished by a continuous pull on the operating lever handle. Figure 18 shows the breechblock operating mechanism. Considering the breech to be closed, the operating lever handle is grasped and pulled to the rear and right. The first part of this pulling movement causes the breechblock to be rotated 1/8 of a turn in the breech recess. This rotation is caused by the action of a rack, located in the breechblock carrier, which engages gear teeth on the top of the breechblock. A lug on the under side of the operating lever operates the rack. Continued

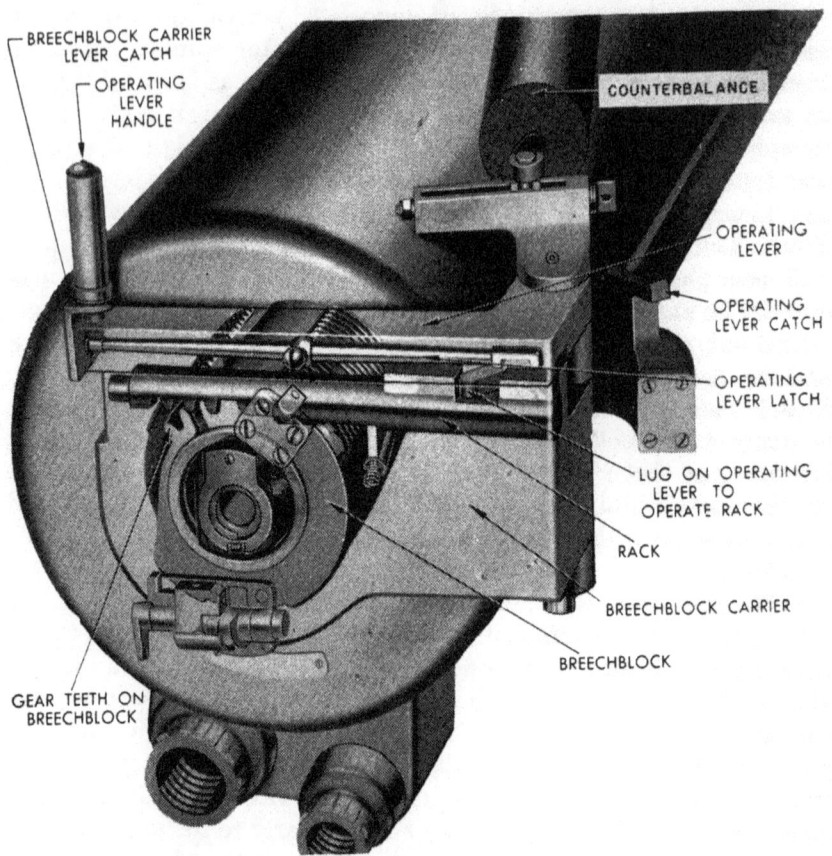

Figure 18. Breechblock, carrier, operating lever, and operating lever handle, 155-mm gun G.P.F., phantom view.

pull on the operating lever handle swings the block carrier to the rear and right, withdrawing the breechblock from the breech recess. The operating lever latch, which runs through the operating lever, locks the breech mechanism in the open position. When closing the breech, the latch is released by a downward pressure on the operating handle and the reverse operation takes place as the operating lever is swung to the left. To facilitate the operation of the breechblock on this gun, when the gun is elevated, a spring-operated counterbalance is provided.

(3) *6-inch gun M1900.* The breechblock mechanism on the 6-inch gun M1900 is similar to that on the 155-mm gun G.P.F., except that the motions in opening (contrary to general practice) are rotation, slight translation, and swing, instead of rotation and swing.

e. STOCKETT BREECHBLOCK. An illustration of the plain slotted-screw, tray-supported type breechblock is the Stockett breech mechanism (fig. 19) which is found on the 12-inch guns. This block is made with six threaded and six slotted sectors. All operations (rotation, translation, and swing) are performed by one crank. The block is opened by turning the crank. The first motion is transmitted through the worm, worm wheel, and compound gear to the rotating lug, which rotates the block the necessary 1/12 of a turn. Further movement of the crank causes the compound gear to engage the translating rack, sliding the block to the rear onto the tray. When the block is sufficiently withdrawn it depresses the rear end of the tray latch, unlocking the tray from the catch on the gun. The further action of the compound gear on the last teeth of the translating rack causes the tray to swing to the right, carrying the block clear of the breech. In recent mechanisms of this type there has been added a locking device which prevents the rotation of the block by the pressure of powder gases.

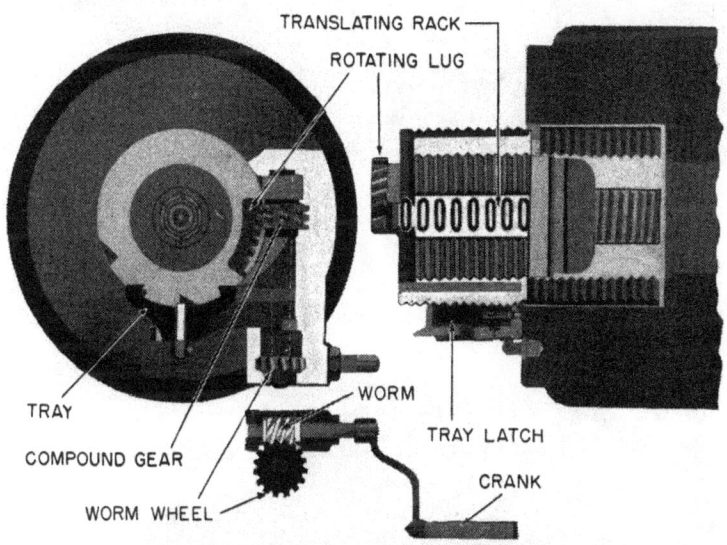

Figure 19. Tray-supported (Stockett design) breechblock mechanism, schematic drawing.

f. WELIN OR STEP-THREADED BREECHBLOCKS. (1) Three examples of the step-threaded form of the slotted-screw block are found in the 155-mm gun M1, the 16-inch gun Mk. II M1, and the 8-inch gun Mk. VI Mod. 3A2. The first two guns utilize two different means of carrier support; the 8-inch gun utilizes a tray support.

(2) *Breechblock, 155-mm gun M1.* This is a carrier-supported type, with threaded sectors of 2 steps, equipped with a spring-actuated counterbalance mechanism (figs. 20 and 21). It has 12 sectors, 4 plain and 8 threaded. To engage or disengage the threads on the

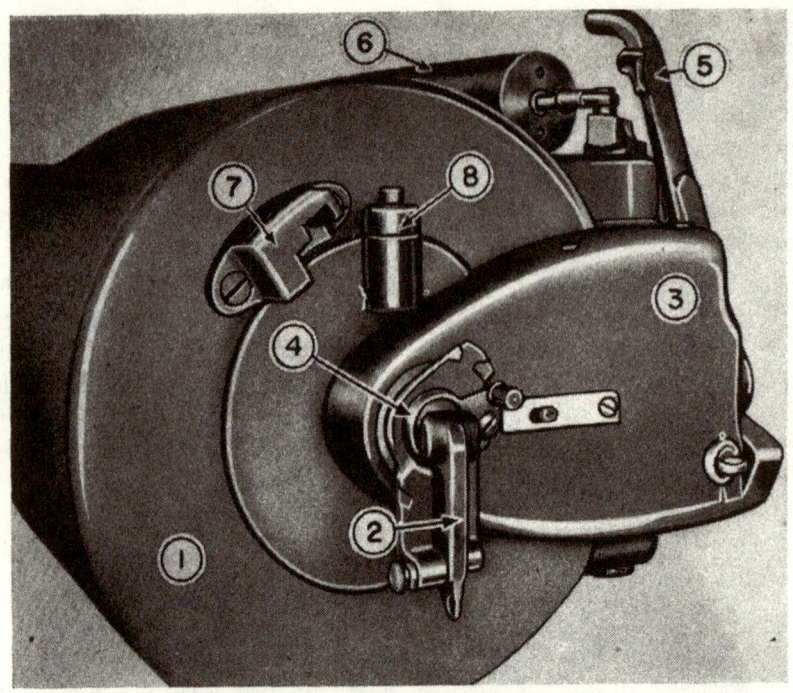

1. Breech ring assembly.
2. Percussion hammer.
3. Breechblock carrier assembly.
4. Firing-mechanism block.
5. Breechblock-operating lever.
6. Counterbalance.
7. Breechblock-rotating cam.
8. Breechblock-rotating roller.

Figure 20. Breech mechanism, closed position, 155-mm gun M1.

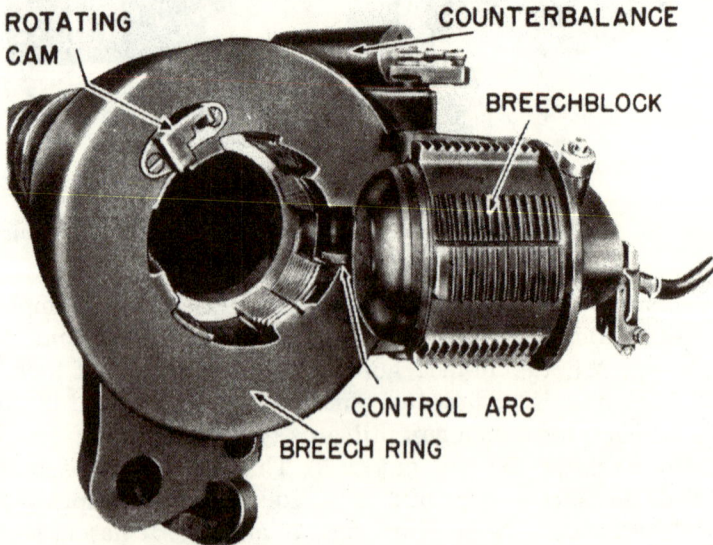

Figure 21. Breech mechanism, open position, 155-mm gun M1.

block with those in the breech recess, the block is rotated 1/12 of a turn. The breechblock is operated by a lever attached to the right side of the carrier. The carrier is hinged to the breech of the gun and supports the breechblock. On the front of the carrier is a pintle, forming a pivot for the breechblock, about which the breechblock rotates when operated. A crosshead (fig. 22) is actuated by means of a crank on the end of a crankshaft, which in turn is operated by the operating lever. The crosshead, operating in a recess in the breechblock, imparts a rotating motion to the block for locking and unlocking operations. A roller on the rear face of the breechblock and a cam on the breech end of the breech ring (fig. 20) are provided to give a turning movement to the breechblock in closing. A control arc (fig. 22) is attached to the pin side of the breech ring face. As

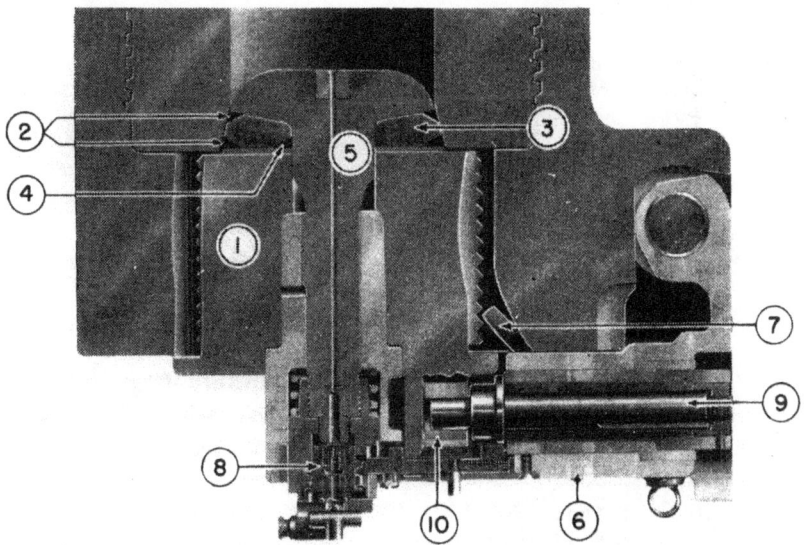

1. Breechblock.
2. Split rings.
3. Gas check pad.
4. Inner ring.
5. Obturator spindle.
6. Breechblock carrier.
7. Breechblock control arc.
8. Firing mechanism assembly.
9. Breechblock-operating crankshaft.
10. Crosshead.

Figure 22. Breech mechanism, 155-mm gun M1, sectional view.

the breechblock is swung into the loading position, it is prevented from rotating by a sector of the block riding over the arc. By its own weight the breechblock is held in an open position. The counterbalance mechanism (fig. 23) compensates for the effect of gravity in the operation of the breech mechanism, thus making it easier to open and close.

25

(3) *Breechblock, 16-inch gun Mk. II M1.* (a) This breechblock (figs. 24 and 25) is of the interrupted step-thread design as previously described, but differs in the number of threaded sectors and type of swing. It contains 12 threaded sectors and 3 plain sectors. The threaded sectors are arranged in 4 steps.

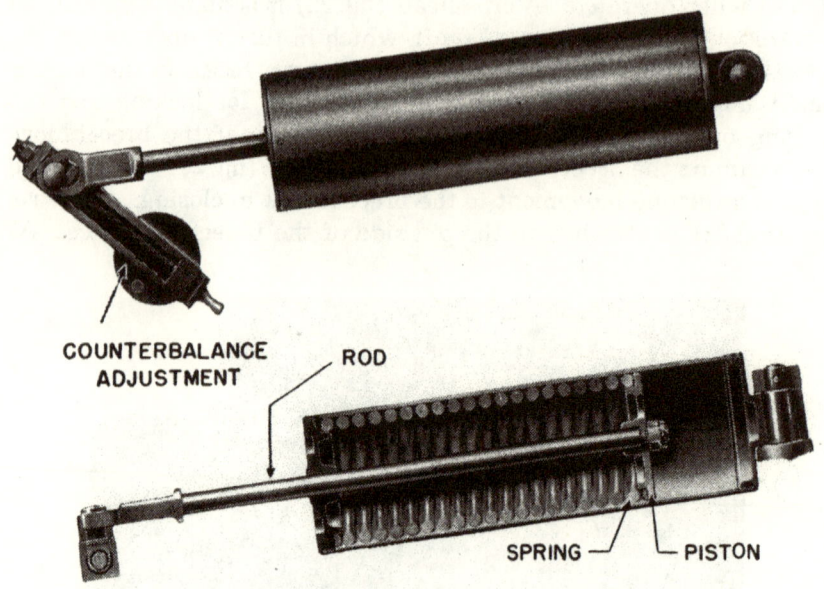

Figure 23. Counterbalance assembly, 155-mm gun M1.

(b) The operation of this breechblock differs from the usual horizontal swing carrier described in the discussion of the breech mechanism of the 155-mm gun M1 in that it has a vertical-swing carrier and is closed by a compressed air device. In opening (fig. 26), a salvo latch is pressed inward allowing the operating lever latch to be raised and thus release the breechblock-operating lever. Pulling on the operating lever activates the connecting rod and rotates the block until the threaded sectors of the breechblock and breech recess are disengaged. Further pull of the operating lever swings the breechblock carrier, which is fastened to the under side of the gun by a carrier hinge pin bracket, in a downward direction until it is open to its maximum limit. Continued swing is prevented by a boss on the lower surface of the carrier coming in contact with the plunger end of a buffer piston. This piston is housed in the carrier buffer attached to the rear end of the recoil cylinder rod.

(c) Two closing cylinders operated by compressed air provide the force for closing the breech. Air is supplied to these cylinders by an air compressor unit which also supplies air for ejecting gases remaining in the tube after firing. It is well to note that this block can be operated by hand if the air compressor system fails. Like other guns of major caliber it is provided with a DeBange-type obturator.

(d) When the breechblock is closed, two cams [upper cam (fig. 40) and lower cam] and two corresponding rollers, attached to the rear surface of the breech and breechblock respectively, automatically initiate a rotary motion to the block causing its threaded sectors

1. Recoil band.
2. Left elevating bracket.
3. Left elevating rack.
4. Firing contactor.
5. Air manifold tubing to recuperator.
6. Air manifold assembly, recuperator system.
7. Left elevating stop.
8. Recuperator cylinder assembly.
9. Firing circuit cable.
10. Recuperator yoke rod.
11. Breechblock.
12. Breechblock carrier.
13. Breechblock-operating lever.
14. Loading platform.
15. Right loading platform beam.
16. Left loading platform beam.

Figure 24. Breechblock, closed position, 16-inch gun Mk. II M1.

to engage those of the breech recess. The completion of this rotary motion is automatically accomplished when the breechblock-operating lever is latched in the closed position. This action is brought about because the operating lever activates the connecting rod which in turn rotates the breechblock.

(4) *Breechblock, 8-inch gun Mk. VI Mod. 3A2.* The 8-inch gun Mk. VI Mod. 3A2, mounted on barbette and railway carriages, provides an example of a step-threaded, tray-supported breechblock (fig. 29). This breechblock contains sectors arranged in three groups, each group consisting of three step-threaded sectors and one plain

Figure 25. Breechblock, open position, 16-inch gun Mk. II M1.

sector. In closing the breech, a 1/12 or 30° revolution of the block is necessary to engage the threads in the breech recess. By turning the operating crank, three motions are given to the breech: rotation, to unlock it; translation, to pull it out of the gun onto the breechblock tray; and swing, to clear the breech of the tray and block. The tray is held in position against the breech by a latch when the breech is

1. Salvo latch and upper rotating cam assembly.
2. Breechblock.
3. Breechblock-operating lever.
4. Connecting rod.
5. Firing-lock retracting lever.
6. Firing-lock operating bar.
7. Operating-lever latch.

Figure 26. Unlatching the breechblock-operating lever, 16-inch gun Mk. II M1.

closed. The latch also prevents the breechblock from sliding off the tray when the breech is open. Approximately four revolutions of the breech operating crank are required to open or close the breech, and the three motions are performed in a continuous operation (figs. 27, 28, and 29).

g. BREECHBLOCKS, 6-INCH GUNS M1903A2 AND M1905A2. The ogival (Bofors) and tapered breechblocks, used on some of the 6-inch guns, are another means of obtaining a larger threaded area and at the same time permitting the block to be shortened. The ogival block (fig. 30), used on the M1903A2 gun, has six slotted and six threaded

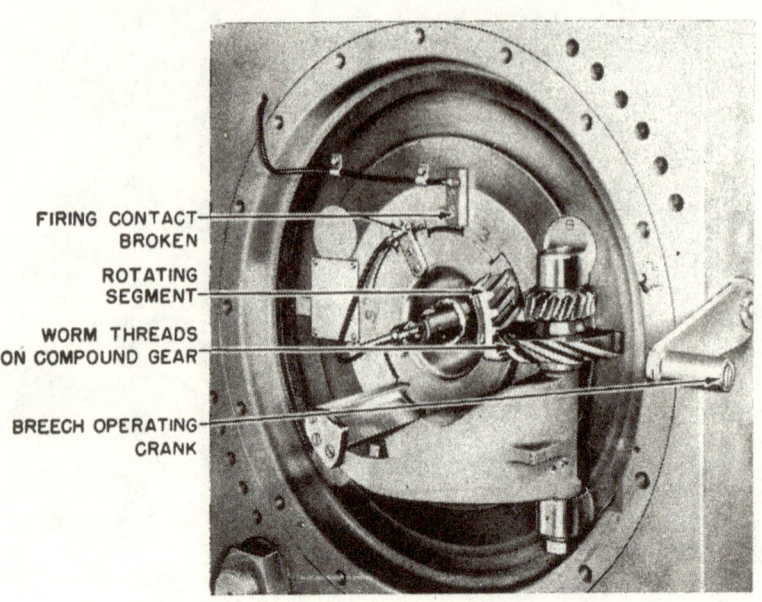

Figure 27. Breech mechanism with breechblock unlocked (rotation), 8-inch gun Mk. VI Mod. 3A2.

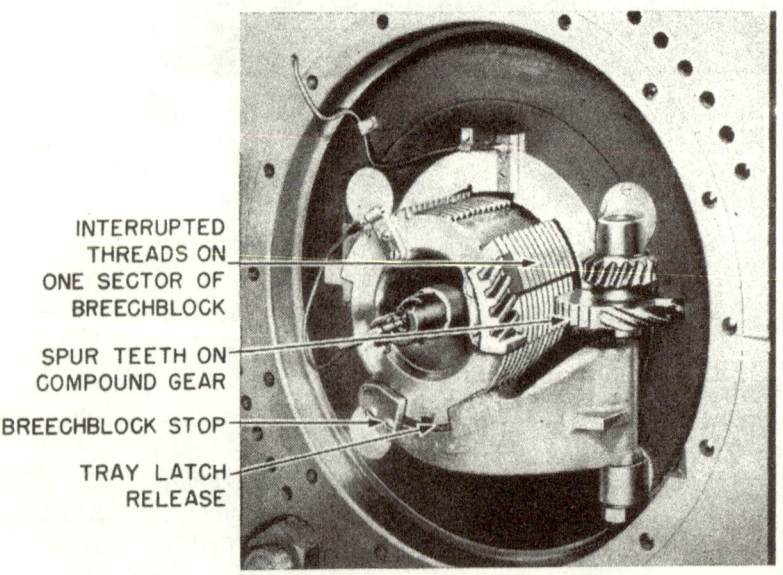

Figure 28. Breechblock on tray (translation), 8-inch gun Mk. VI Mod. 3A2.

Figure 29. Breechblock and tray swung clear of the breech recess (swing), 8-inch gun Mk. VI Mod. 3A2.

segments. Because of its shape and slotted sectors, only a small retraction to the rear is necessary to swing the block open. The tapered block (fig. 31), employed on the M1905A2 gun, has six slotted and six threaded segments which facilitate its opening and closing in the same manner as the ogival block. Both types of breechblocks are carrier-supported, swinging to the right on a hinge mechanism attached to the right side of the breech. The breech mechanism for both model guns is of the lever-pull type. Two motions of the breechblock, rotation and swing, are involved in opening and closing.

15. Sliding-wedge Breechblock, 90-mm Gun

a. GENERAL. The sliding-wedge breechblock is well adapted for rapid-fire weapons with automatic methods of operation using fixed ammunition (par. 13). It is not used in our service with separate-loading ammunition. It necessitates the use of a comparatively large breech section to withstand the stresses of firing, thus adding considerably to the weight of the gun. The mechanism employs a rectangular wedge-shaped block, securely seated in a slot cut in the breech of the gun, perpendicular to the bore, and operated by a hand or automatic crank or lever device. The motion of the block may be either horizontal or vertical. The latter is generally classified as the drop-block type. The 90-mm gun M1 (fig. 32) provides an excellent example of the use of this type of block. The design of the breechblock permits either automatic or manual operation.

b. DESCRIPTION. The breechblock (fig. 33) has a guide rib on each side which slides in a corresponding groove in the side of the breech recess when the block is raised or lowered. The top of the block is U-shaped to guide the cartridge (in loading) into the chamber, while the upper front edge is beveled to drive the cartridge into the chamber as the breechblock is raised. The rear face and guides of the breechblock and the rear wall of the breech recess are inclined so that the breechblock, in rising, moves forward to complete the seating of the cartridge as the breech is closed. In each

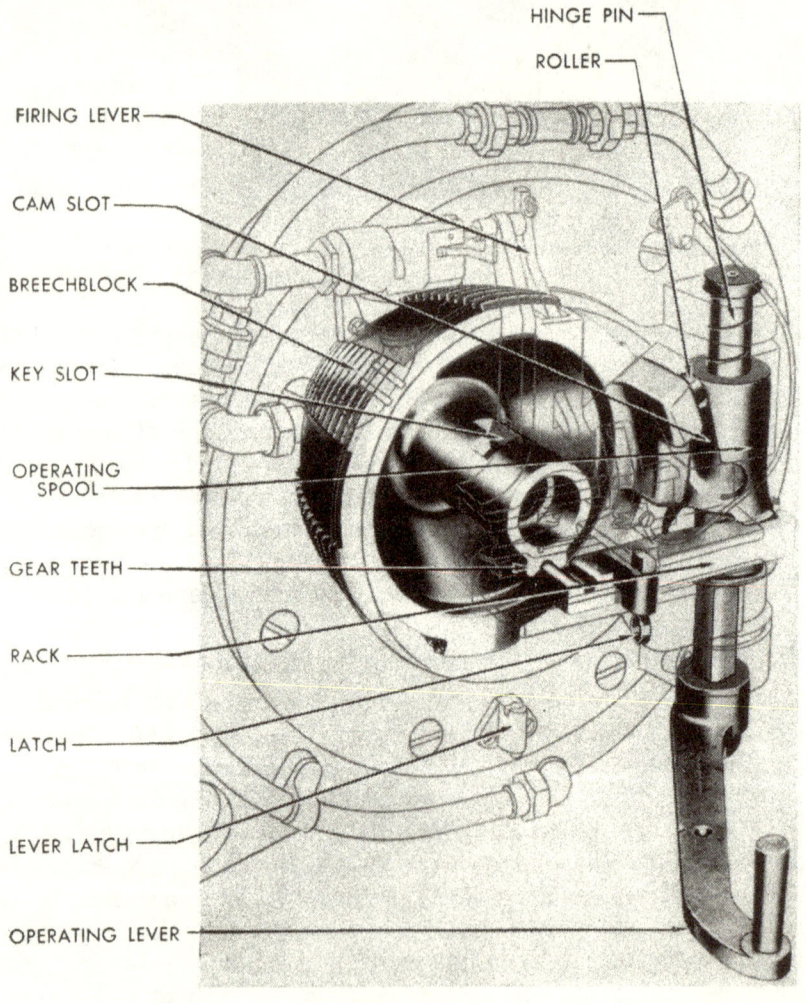

Figure 30. Breech mechanism, 6-inch gun M1903A2.

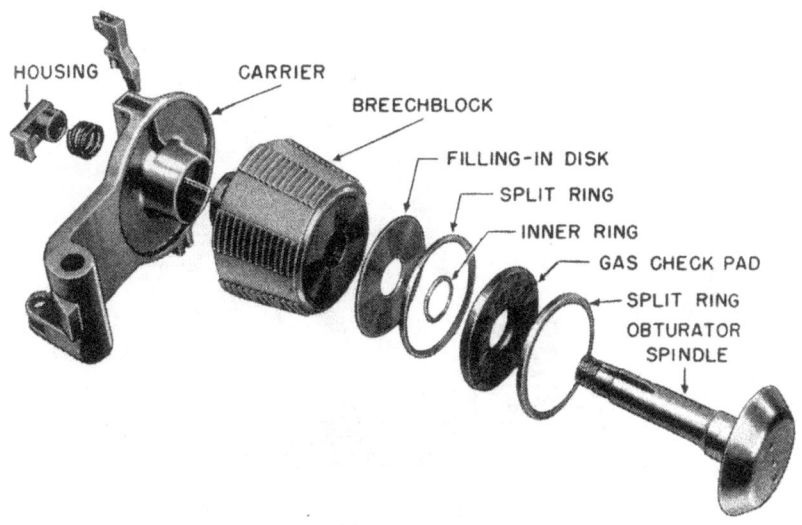

Figure 31. Breech mechanism, 6-inch gun M1905A2, exploded view.

side of the breechblock is a groove (fig. 37) in which the inner trunnions of the extractors slide. The bottom of the breechblock contains an inclined T-slot, in which the crosshead (fig. 34) of the breechblock crank slides to raise and lower the block. The breechblock houses the firing mechanism and allied parts (fig. 33).

c. OPERATION. The breechblock is raised or lowered by the rotation of the spline shaft (fig. 34) which passes transversely completely through the lower part of the breech ring (fig. 35). On this shaft are three cranks. The center or breechblock crank raises and

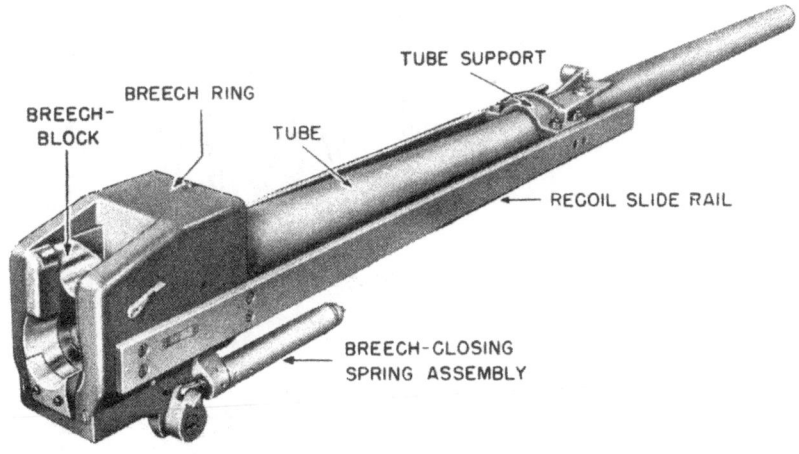

Figure 32. Gun assembly with sliding-wedge breechblock, 90-mm gun M1.

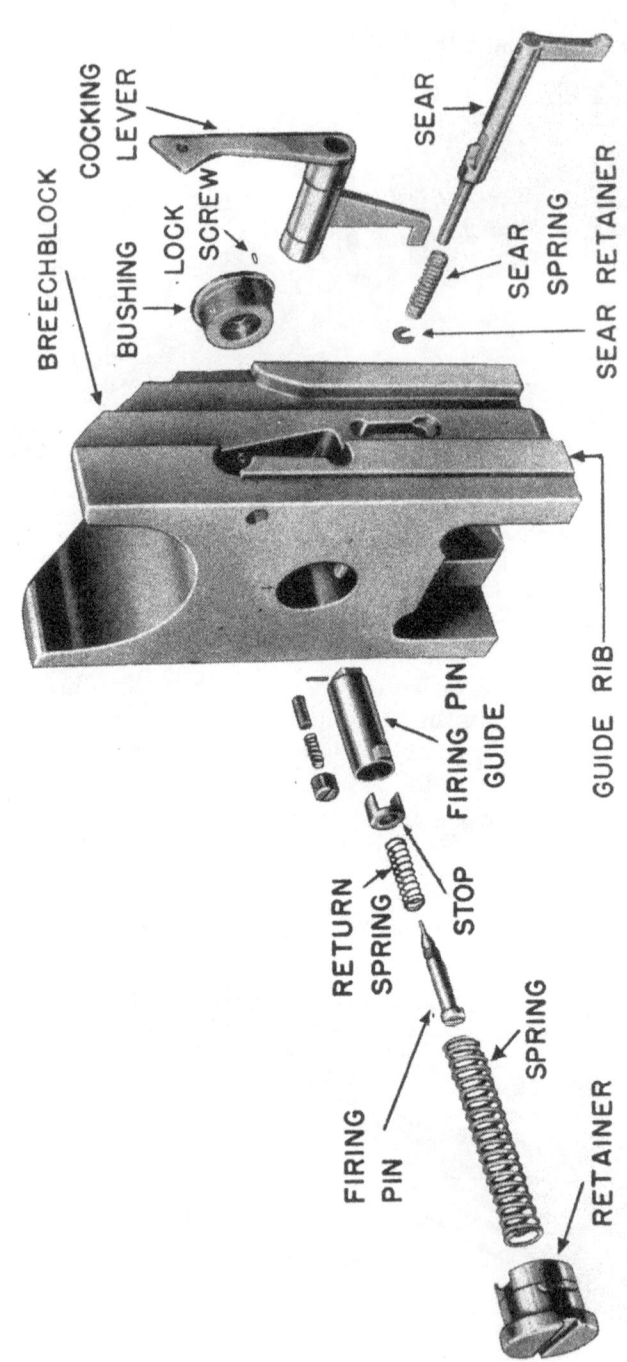

Figure 33. Breechblock mechanism, 90-mm gun M1, exploded view.

lowers the breechblock; the other two (operating crank and chain terminal crank) operate the center crank as follows:

(1) *Automatic.* When a breech-operating cam lever (not shown) on the left side of the cradle is swung up and latched in the automatic position, the breech automatically opens after each round is fired. As the gun slides forward in counterrecoil, the operating crank (fig. 35) engages the breech-operating cam (not shown) and the spline shaft is rotated. This causes the arm of the breechblock crank to swing rearward and downward, thus lowering the block. Rotation of the spline shaft also causes the spring in the closing spring cylinder (fig. 36) to be compressed. The extractors (fig. 37) lock the breech in the open position. When the extractors are activated by the ramming of a cartridge into the firing chamber, the released, compressed closing spring causes the chain terminal crank

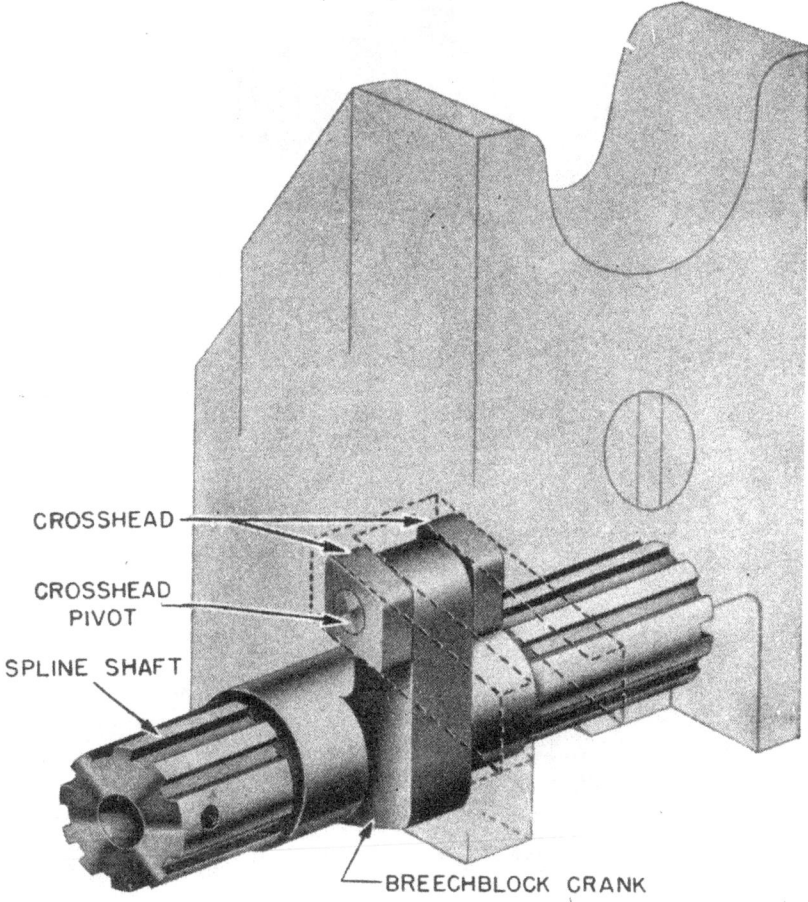

Figure 34. Breechblock crank and spline shaft, 90-mm gun M1.

(figs. 35 and 36) to rotate, thus returning the breechblock to raised position.

(2) *Manual.* When the breech-operating cam lever is swung down and latched in the hand position, opening of the breech is accomplished by pulling back on the breech-operating handle (not shown) located on the right side of the cradle. The gun must not be over ½ inch out of battery to permit this operation.

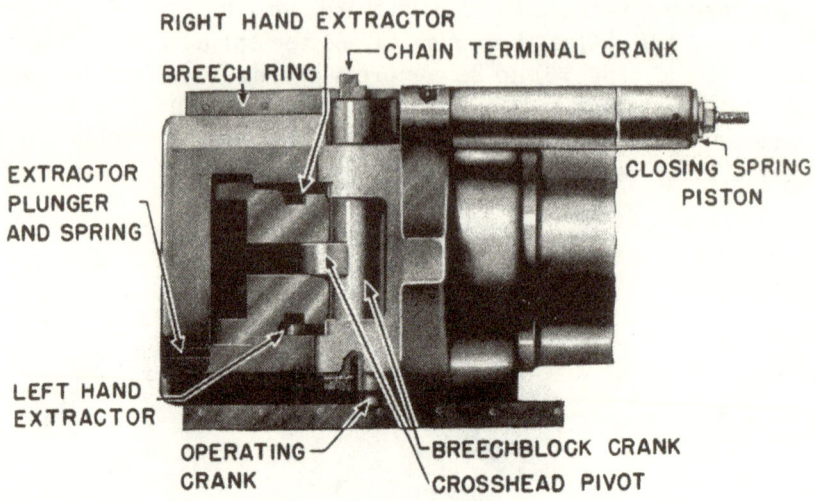

Figure 35. Breech mechanism, bottom view, 90-mm gun M1.

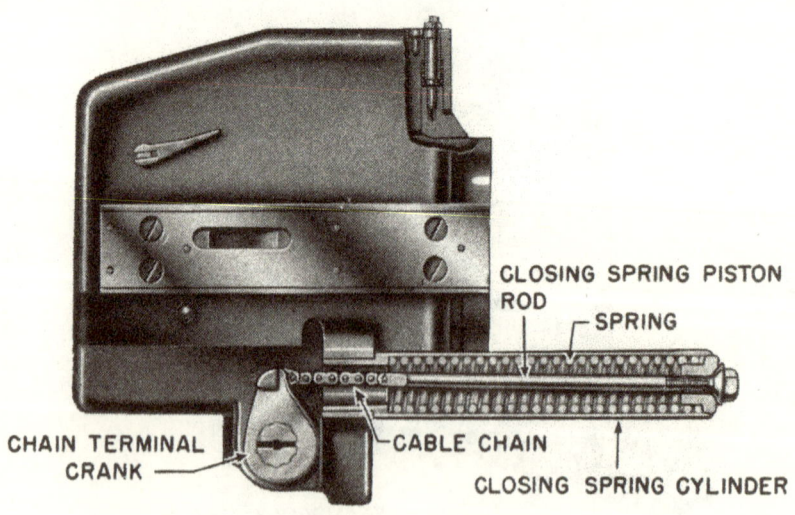

Figure 36. Breech mechanism, right side, 90-mm gun M1.

d. ACTION OF THE EXTRACTORS. The cartridge case extractors (fig. 37) fit between the side of the breechblock and the breech ring. As the breechblock moves downward, the extractors are automatically activated by the forward curves of the breechblock grooves,

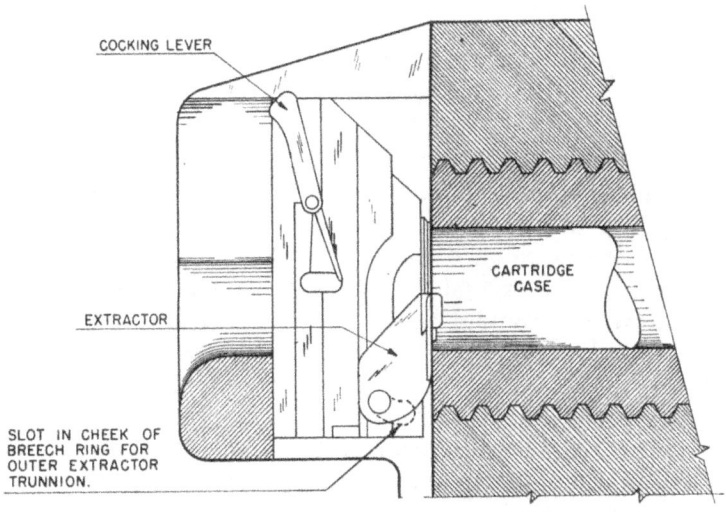

A. EXTRACTORS WITH BREECHBLOCK IN CLOSED POSITION

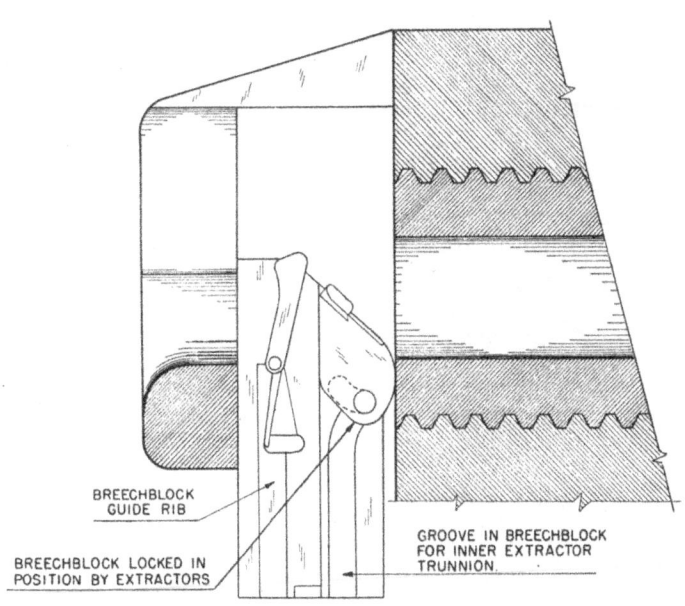

B. EXTRACTORS WITH BREECHBLOCK IN OPEN POSITION

Figure 37. Extractors, 90-mm gun M1.

the cartridge is uncovered, and the lips on the upper inner edge of the extractors engage the rim of the cartridge case and move rearward, drawing the case out of the chamber. As the breechblock reaches the full open position, spring-activated extractor plungers (fig. 35) force the inner trunnions of the extractors forward onto the lower, horizontal flat edges of the breechblock grooves, locking the breechblock in the open position.

16. Pressure Gages

a. Pressure gages for most cannon using separate-loading ammunition are mounted in receptacles provided for them on the mushroom head of the obturator spindle. The two large nuts found on the mushroom head are dummy pressure plugs. When it is desired to use pressure gages during firing, these dummies are removed and similar devices containing the real pressure gages are installed.

b. The pressure gage consists essentially of a copper cylinder suitably mounted. When this cylinder is subjected to the powder pressure, it is deformed in proportion to the pressure exerted. By measuring the cylinder before and after firing, it is possible to tell the maximum pressure developed in the powder chamber.

c. In those cannon not equipped for the use of the gages mentioned above, it is necessary to place gages loosely in the powder chamber and retrieve them after firing the round and before another round is inserted.

Section III. FIRING MECHANISMS

17. General

This section considers the basic fundamentals of the operation and construction of firing mechanisms and discusses the types most commonly employed in seacoast artillery.

18. Firing Lock Mk. I

a. GENERAL. (1) This type of firing mechanism is used on the 16-inch gun Mk. II M1 as well as on other larger caliber weapons such as the latest 14-inch railway guns. The primer (fig. 38) used is a combination electric-percussion type fired either by electricity or by lanyard.

(2) The assembled firing lock (fig. 40) is attached to the end of the obturator spindle by merely giving it a ¼ turn to lock it in position since the end of the spindle and the firing-lock housing are threaded similarly to the slotted-screw type breech mechanism (par. 14*b*).

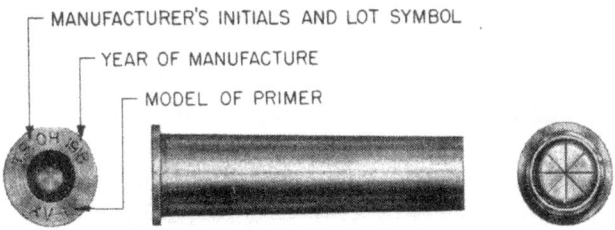

Figure 38. Primer, combination, electric and percussion, Mk. XV Mod. 1.

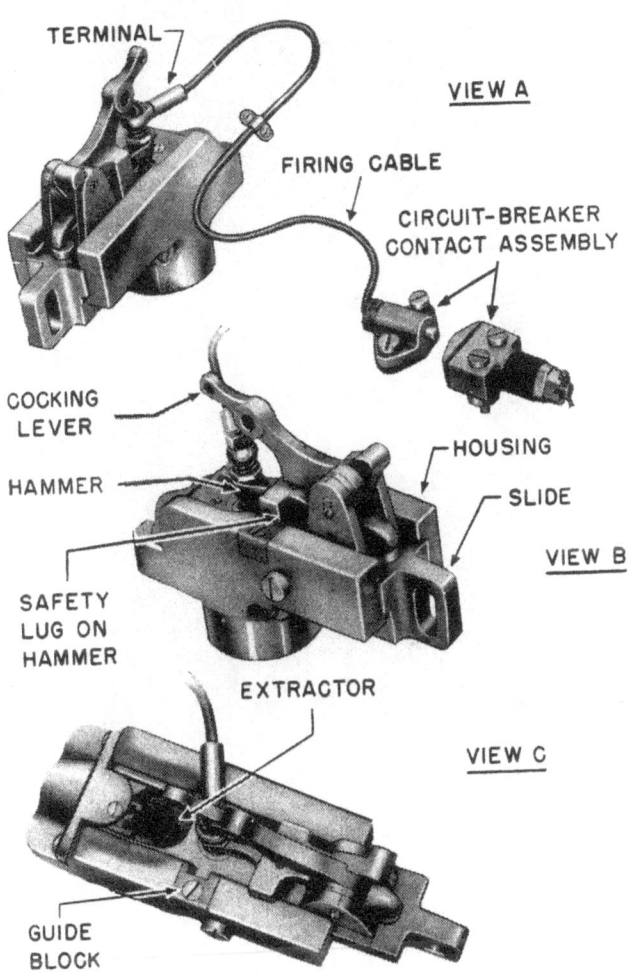

Figure 39. Firing lock Mk. I, Navy type.

b. OPERATION. (1) The principal parts of the lock are shown in figures 39 and 40. They are the housing, slide, firing-lock operating bar, cocking lever, hammer, and extractor unit. A primer-retainer catch with its operating spring is housed in the upper part of the firing lock. The extractor, a fork-shaped piece which straddles the head of the primer, is actuated by an extractor cam provided with a torsional spring. The operating bar and slide move up and down as a unit; the slide moves in grooves in the sides of the firing-lock

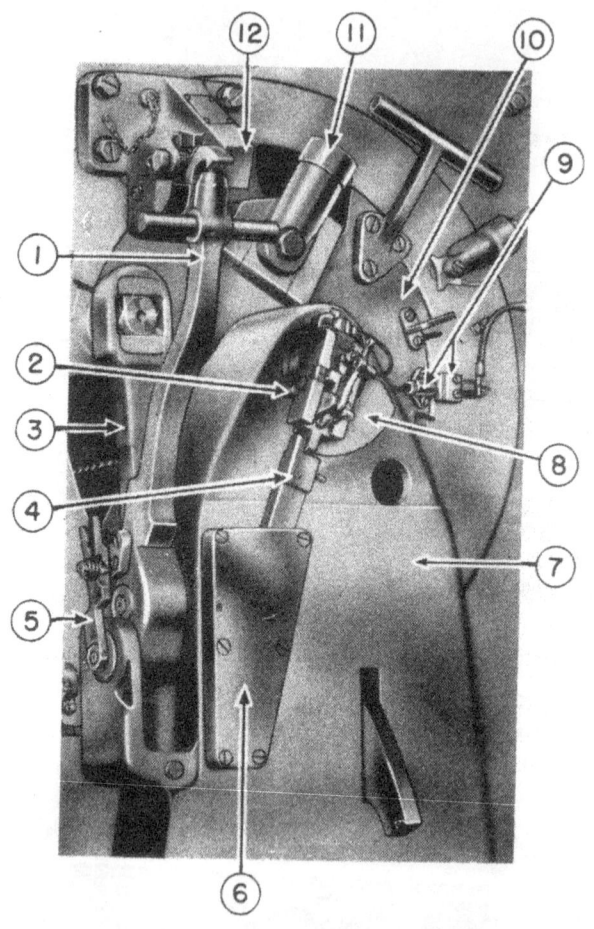

1. Breechblock-operating lever.
2. Firing-lock assembly.
3. Connecting rod.
4. Firing-lock operating bar.
5. Firing-lock retracting lever.
6. Operating-bar bearing plate.
7. Breechblock carrier.
8. Firing-lock safety arc.
9. Circuit-breaker contact assembly.
10. Breechblock.
11. Upper rotating roller.
12. Upper rotating cam.

Figure 40. Breechblock and firing lock Mk. I, operating parts, 16-inch gun Mk. II M1.

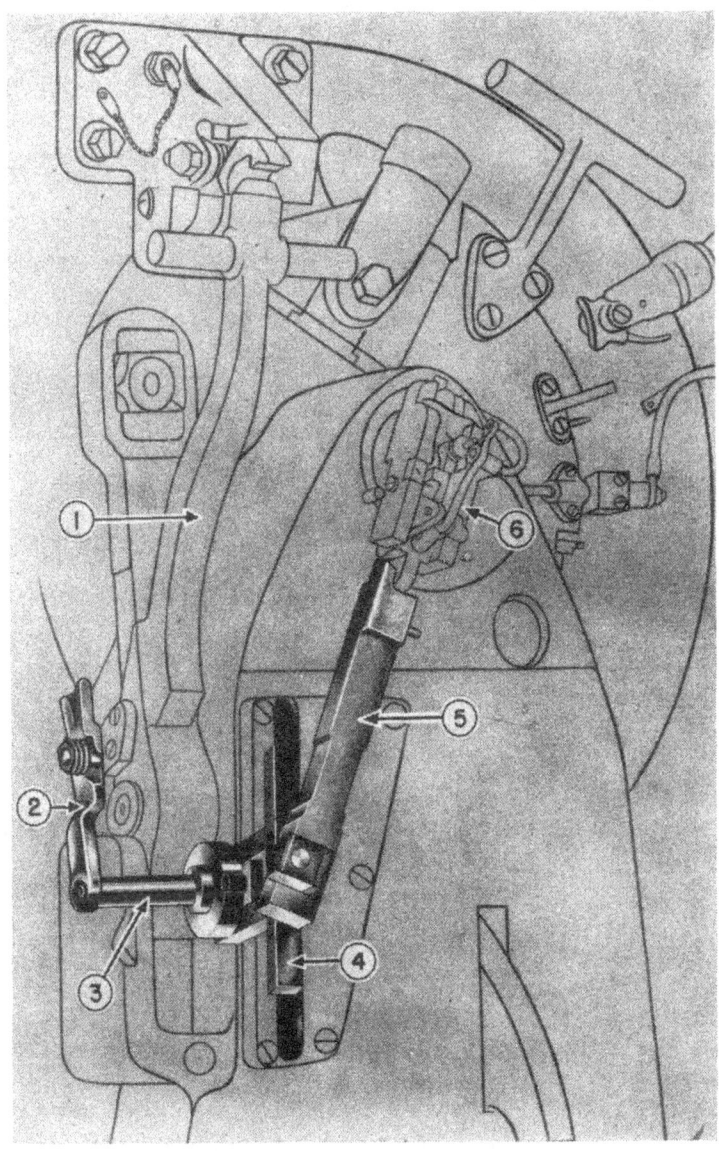

1. Breechblock-operating lever.
2. Retracting lever.
3. Retracting-lever crank.
4. Firing-mechanism crosshead.
5. Operating bar.
6. Firing lock.

Figure 41. Operating mechanism for firing lock Mk. I, 16-inch gun Mk. II M1. Drawing of mechanism superimposed on relative position in breechblock carrier.

housing and the operating bar moves in a slot in the bearing plate on the rear face of the breechblock carrier. A hardened faceplate is placed in front of the slide to take the thrust of the primer when fired.

(2) When the breechblock is opened, the retracting-lever crank (fig. 41), manipulated by the breechblock-operating lever through the retracting lever, pulls down on the firing-lock operating bar which lowers the slide until the primer seat is uncovered. A catch on the breechblock-operating lever causes the retracting lever to be lowered when the breechblock-operating lever is moved. The retracting lever may also be operated by hand, independent of the

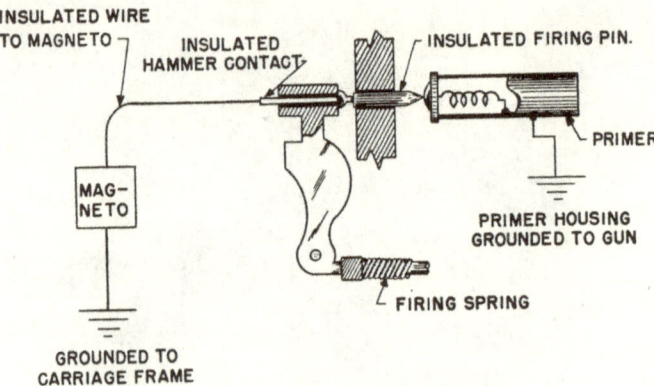

A. ELECTRIC FIRING

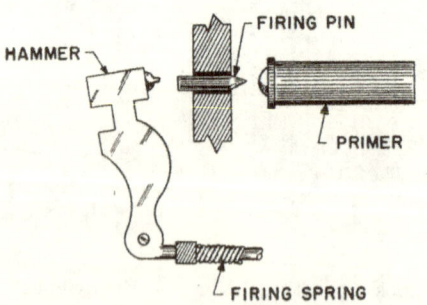

HAMMER DRAWN TO REAR – WHEN TRIPPED, FIRING SPRING WILL FORCE HAMMER FORWARD – FIRING PRIMER.

B. LANYARD FIRING

Figure 42. Firing lock Mk. I, operation, schematic drawing.

breechblock-operating lever. As the slide of the firing lock is lowered, it works against a small extractor cam, rotating the extractor to the rear and ejecting the fired primer automatically.

(3) When the breechblock is closed, the firing-lock operating bar is automatically raised, which in turn raises the slide until the firing pin is opposite the primer and the lock is ready to fire. For safety reasons, the primer is not inserted until the block is fully closed. Thus, after the block is closed, the operating bar (fig. 41) is pulled down by hand (using the retracting lever), the primer inserted, and the operating bar raised to firing position.

(4) For percussion firing, the lanyard is inserted in the hole in the cocking lever (fig. 39). The first part of the lanyard's travel cocks the hammer, while the last part trips the hammer, allowing it to fly forward and strike the firing pin as shown in B, figure 42. The action is identical with that of the double-action revolver.

(5) For electric firing, a current of electricity is sent through the insulated hammer and the insulated firing pin to the cap of the primer as shown in A, figure 42. Current for firing the piece may come from an outside source of power (not shown) or from one of two magnetos located on the right and left sides of the carriage. The magnetos are operated independently to produce a firing spark. The magnetos are *never* operated simultaneously. When the gun is fired by means of the outside power source, a transformer reduces the line voltage, and a push-button system enables the gun commander to fire the gun from his position in the rear. A circuit-breaker contact (figs. 39 and 40) is attached to the gun breech and, when in contact with the circuit breaker on the breechblock, provides a means for an uninterrupted flow of current to the primer in the firing lock.

c. SAFETY FEATURES. This lock possesses the following safety features:

(1) The slide is held down as long as the breechblock is open and does not allow the firing pin to move up into firing position opposite the primer until the block is completely closed.

(2) Only when the breechblock is fully closed is the safety lug on the hammer in line with an opening on the guide block (B and C, fig. 39). When in this position, the hammer may be operated.

(3) A firing-lock safety arc (fig. 40), attached to the upper end of the breechblock carrier, limits the movement of the firing lock and prevents firing of the piece until the lock is properly assembled to the end of the obturator spindle.

(4) The circuit breaker on the breechblock is in contact with the circuit-breaker contact on the gun breech only when the breechblock is completely closed.

43

(5) A recoil firing contactor (fig. 24), attached to the cradle and recoil band, prevents firing of the piece except when in battery position.

19. Seacoast Firing Mechanism M1903

a. GENERAL. (1) Firing mechanism M1903 (fig. 43) is used on practically all seacoast cannon manufactured prior to the European War, 1914-1918. It is designed to fire by lanyard (using a friction primer) or by electricity (using an electric primer). The principal parts of the firing mechanism are: hinged collar, slide, firing leaf, safety bar, ejector, and handle and slide catch.

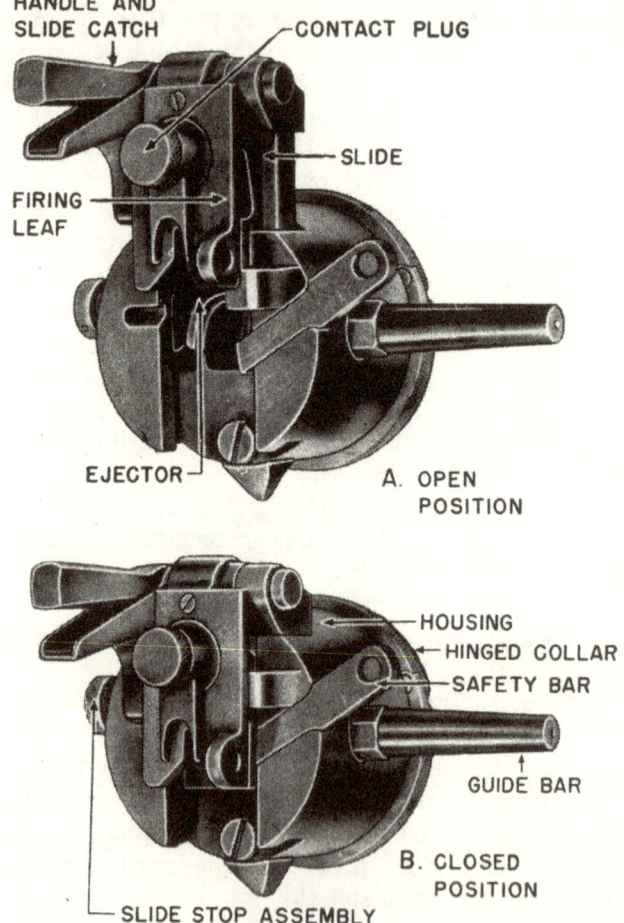

Figure 43. Seacoast firing mechanism M1903, assembled view, open and closed position.

(2) A seat for the firing mechanism is provided on the rear end of the obturator spindle. The hinged collar is attached to the end of the spindle by means of two grooves which engage corresponding ribs on the spindle. The collar is threaded on the outside to receive the firing mechanism housing which is locked to the collar by a spring pin. In placing the firing mechanism, the housing is held stationary while the hinged collar is turned. The firing mechanism is free to turn on the spindle as the breechblock is rotated.

b. OPERATION. (1) As shown in A, figure 44, the primer resembles a blank rifle cartridge with a wire protruding from the rim end. Referring to B, a firing magneto (or outside power source with transformer) is used in electrical firing. One end of the line is connected by an insulated wire to the wire part of the primer, and the other end grounded to the frame of the gun. When using the friction primer, the lanyard is hooked to the lower end of the firing leaf and, when pulled, draws the wire to the rear, igniting the primer (C, fig. 44).

A. PRIMER—FRICTION OR ELECTRIC

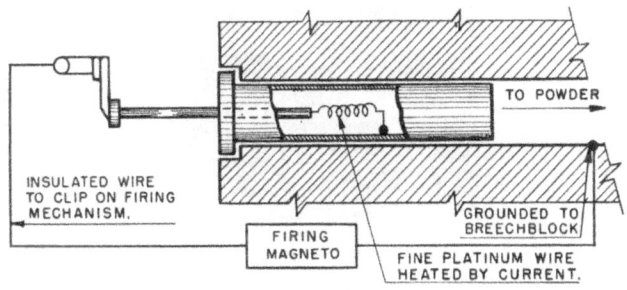

B. HOOKUP OF ELECTRIC FIRING DEVICE

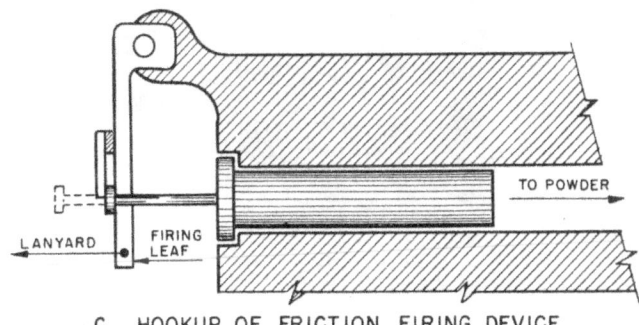

C. HOOKUP OF FRICTION FIRING DEVICE

Figure 44. Seacoast firing mechanism M1903, operation, schematic drawing.

(2) The ejector, pivoted in the housing, has at its lower end a forked seat for the primer head. When the slide is raised, a shoulder strikes the upper end of the ejector and the lower end of the latter is thrown to the rear, ejecting the fired primer by means of the lip on the lower end.

(3) The housing carries projecting ribs, forming guides for the slide which is moved up and down by the handle after the slide catch is released. At its upper end, the firing leaf is pivoted to the

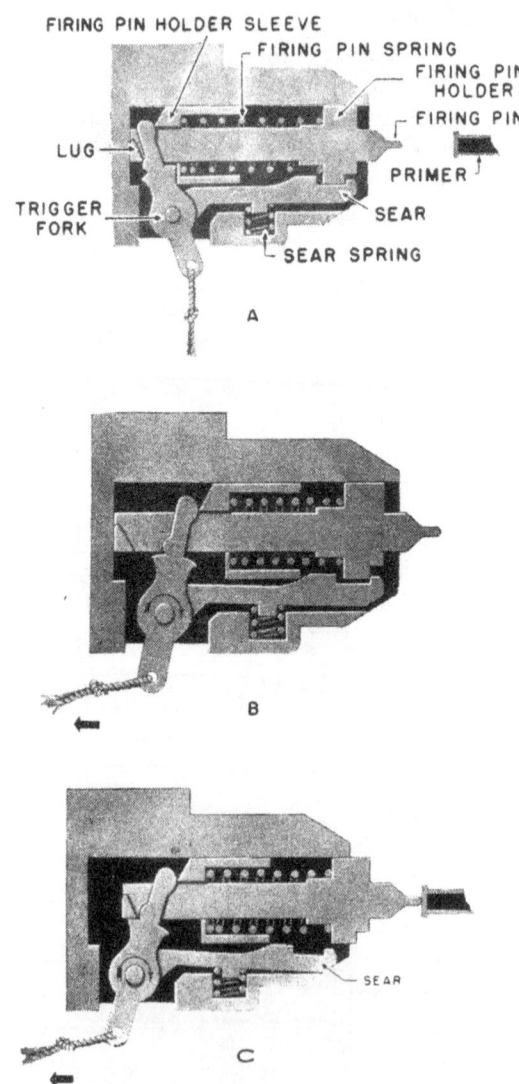

Figure 45. Operation of continuous-pull firing mechanism, schematic drawing.

slide. It has a vertical slot in the lower edge through which the wire of the primer projects when the slide is in its lower or locked position. If the breechblock is closed, a pull on the lanyard rotates the firing leaf about its axis, drawing out the primer wire and firing the primer. A closing of the electric circuit, which enters the mechanism through the electric terminal, will fire the primer by electricity. In inserting the primer, it is necessary to start the slide downward in order to engage the ejector lip behind the rim of the primer.

c. SAFETY FEATURES. This mechanism has the following safety features:

(1) The firing leaf cannot be drawn back for lanyard firing (because the safety bar engages the leaf) until the block is completely closed.

(2) Because of a circuit breaker, the firing circuit is not made at the breech until the block is completely closed.

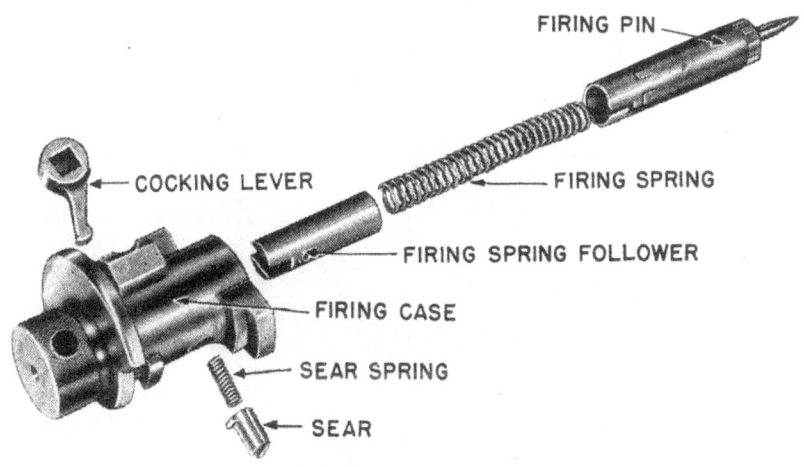

Figure 46. Firing mechanism, 3-inch gun M1903, exploded view.

20. Continuous-pull Type Firing Mechanism

a. GENERAL. The continuous-pull type of firing mechanism (fig. 46) is used on the 3-inch guns M1902M1 and M1903. Since the propelling charge for these guns is contained in a metallic case, it is necessary to fire a percussion primer in the base of the case. The operation of the mechanism is actuated by one continuous pull. However, there are really three phases in its firing cycle:

(1) *Cocking phase.* The initial part of the pull on the lanyard compresses the firing pin spring (B, fig. 45).

(2) *Firing phase.* The remaining part of the lanyard's movement disengages the sear, enabling the spring to expand and force the firing pin against the primer, firing it (C, fig. 45).

(3) *Retracting phase.* Here the recoil of the gun allows the lanyard to slack, thus permitting the trigger fork (A, fig. 45) to rotate to the rear and to bring the firing spring and pin with it. This prevents injury to a protruding firing pin and prevents the scoring of cartridge cases when the block is opened. After the primer has been struck, the firing pin spring is still under compression and

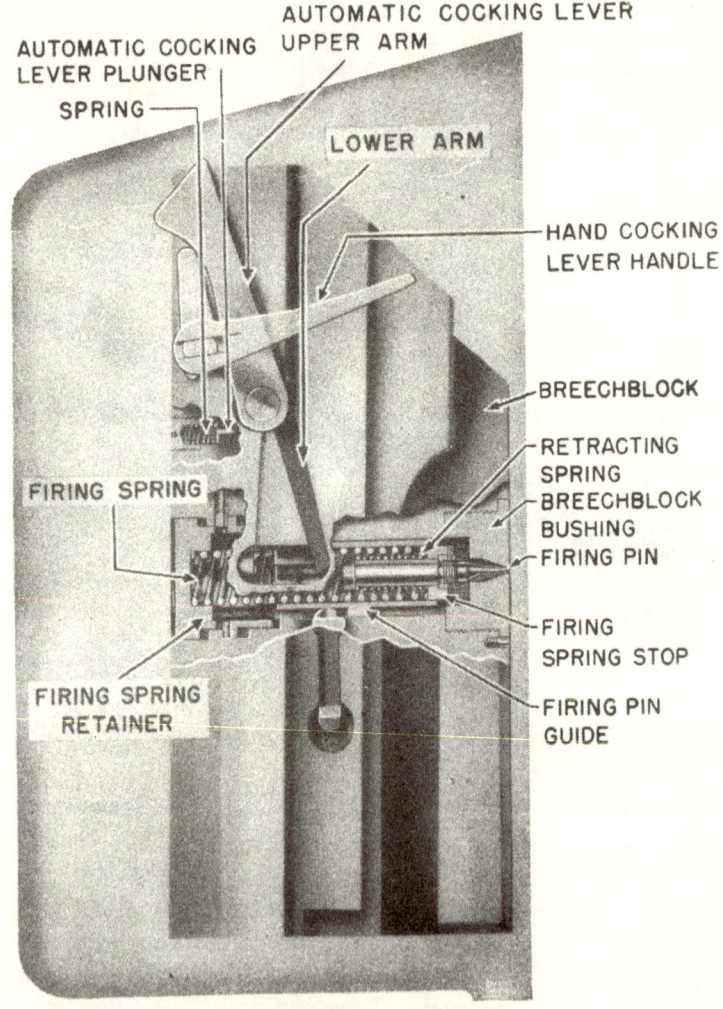

Figure 47. Firing mechanism, 90-mm gun M1.

hence tends to expand. As its forward movement is prevented by the firing pin holder, it moves to the rear, rotating the trigger fork.

b. SAFETY FEATURES. This mechanism cannot operate by a sudden jar as the firing pin spring cannot be compressed except by a pull on the lanyard. Mounted in a drop block (par. 15), the firing pin does not appear opposite the cap in the cartridge until the breechblock is completely closed.

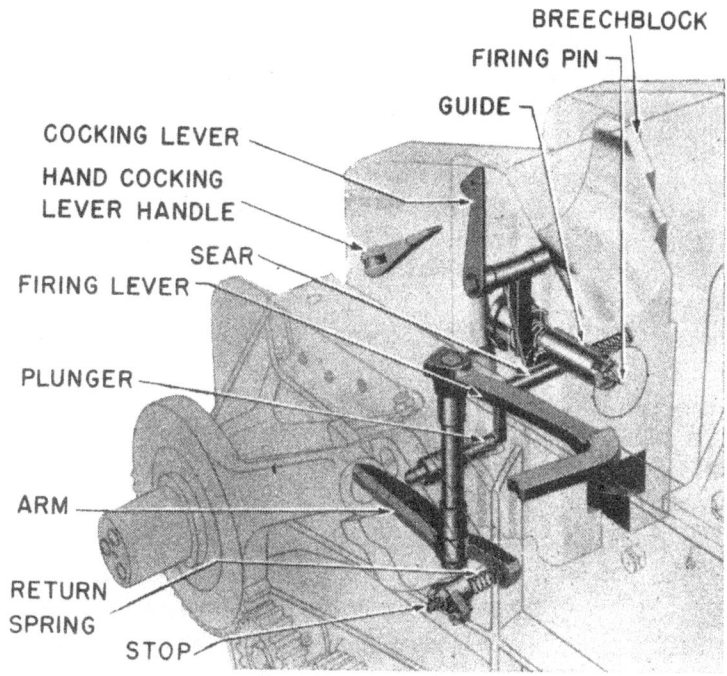

Figure 48. Train of parts which operate firing mechanism, 90-mm gun M1.

21. Inertia-type Firing Mechanism

The inertia type is a modification of the continuous-pull mechanism; it is well illustrated by the firing mechanism on a 90-mm gun M1 (figs. 33, 47, and 48). Like the continuous-pull firing mechanism, its operation is divided into three phases:

a. COCKING PHASE. This phase of the action is accomplished by an automatic cocking lever (fig. 33). The rotation of the automatic cocking lever, caused by the dropping of the breechblock into the open position, results in a rearward motion of the lower arm of the cocking lever (fig. 47). The lower arm engages a cocking lug and pulls the firing pin guide to the rear, thus compressing the firing spring. When the hand cocking lever is used, it is rotated forward to force the upper arm of the automatic cocking lever to return the firing mechanism to the cocked position without the breech being opened.

b. FIRING PHASE. (1) When the firing mechanism is in cocked position, the sear, which passes through the breechblock below the guide (figs. 33 and 48), engages the sear lug in the bottom of the guide, preventing the guide from moving forward. When the plunger is pressed inward, the lug slips through the slot in the sear, and the firing pin guide and firing pin are released to fly forward under the pressure of the firing spring (fig. 47).

(2) The gun is fired by pulling the firing lever (fig. 48). The motion is transferred down through the side of the cradle by a shaft to which is attached an arm. The arm then presses the plunger to release the sear.

c. RETRACTING PHASE. Just before the firing pin strikes the primer, the action of the firing spring is stopped by the firing spring stop (fig. 47) striking the inside rear surface of the breechblock bushing. The firing pin and guide, however, are carried forward by inertia to strike the primer. This final motion compresses the retracting spring between the firing pin head and the stop. The spring then expands. As it expands, it retracts the firing pin and guide so that the point of the former is flush or slightly to the rear of the front face of the breechblock. Damage to the firing pin or breechblock is in this way prevented.

22. 155-mm Firing Mechanism

a. GENERAL. The 155-mm G.P.F. and the 155-mm M1 firing mechanisms are for the most part the same in design. Each uses percussion primers that resemble blank cartridges for a revolver. The parts of the firing mechanism (figs. 50 and 51) are incased in a block which is also provided with a handle for use in screwing the assembly in or out of the firing mechanism housing (figs. 20 and 49). The housing is screwed to the rear end of the obturator spindle. In addition to the relatively large threads on the outside of the block, used for screwing it in the housing, the firing mechanism block is threaded in its interior to receive the primer holder. The latter has a U-shaped slot for engaging and holding the rim of the primer. The pressure of the compression spring bearing against the firing pin guide holds the primer in place. The pressure of the spring also prevents the firing pin from hitting the primer before it is struck by the percussion hammer (fig. 49). The primer is fired by pulling the lanyard attached to the bottom of the percussion hammer. The latter is pivoted on a lug projecting from the breechblock carrier.

b. OPERATION. Each time the breechblock is closed on a loaded round, the firing mechanism block is screwed into the firing mechanism housing. A firing mechanism block latch is attached to the carrier (on the G.P.F.) at the right and a little above the firing mechanism housing. Its function is to prevent the mechanism from unscrewing during firing. After the gun is fired and before the breech is opened the following steps are taken: the latch is pressed back to free the handle of the firing mechanism block, the block is unscrewed, the used primer is taken out of the slot in the primer holder and a new primer inserted. After the breech is completely closed, the block is again screwed into the firing mechanism housing. The latching of the M1 firing mechanism (fig. 20) is somewhat different from that of the G.P.F. but the principle of operation is essentially the same.

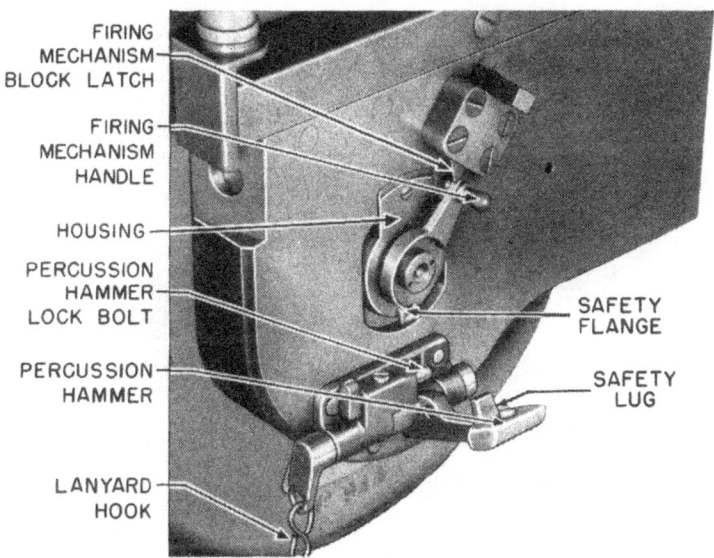

Figure 49. Firing mechanism screwed home, percussion mechanism, and block latch, 155-mm gun G.P.F.

c. SAFETY FEATURES. (1) As a safety feature, there is a flange (figs. 49 and 51) on the firing mechanism block in which is cut a recess; there is also a safety lug on the front of the percussion hammer. The flange and lug prevent the hammer from striking the firing pin except when the firing mechanism block is fully inserted and the lug and recess are lined up. However, this is defective inasmuch as the primer may be fired when the block is not screwed completely home; therefore, it is imperative that no attempt be made to insert the firing mechanism block until the breechblock is fully closed.

(2) The firing mechanism housing is equipped with a safety plunger (not shown) which prevents the complete insertion of the firing mechanism until the breechblock is fully closed.

(3) A firing mechanism block latch (fig. 49) prevents the firing mechanism from unscrewing during firing.

(4) A percussion hammer lock bolt (fig. 49), which is intended to hold the percussion hammer stationary when the gun is in traveling position and out of action, may also be used as a safety precaution. In firing, the bolt is locked immediately after the breech is opened and it is not unlocked until after the breechblock is completely closed and the gun is ready to be fired.

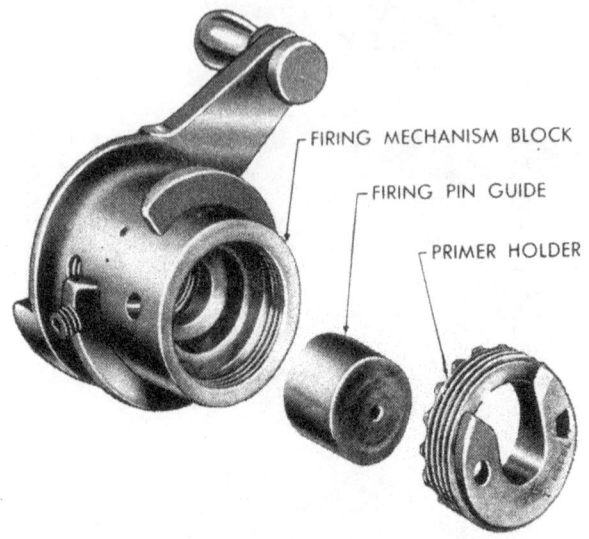

Figure 50. Forward end of firing mechanism, 155-mm gun G.P.F., exploded view.

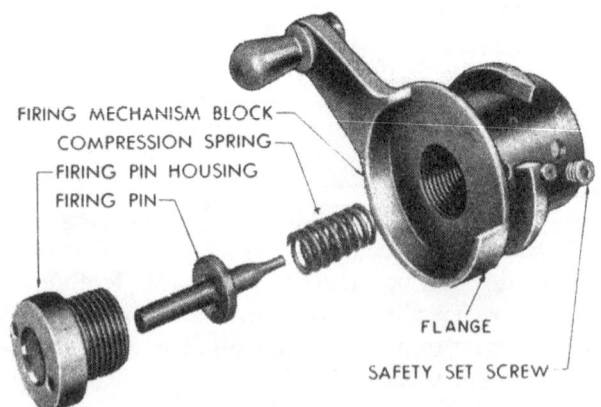

Figure 51. Rear end of firing mechanism, 155-mm gun G.P.F., exploded view.

23. Firing Lock Mk. VIII Mod. II

a. The firing lock Mk. VIII Mod. II (figs. 52, 53, and 54) used on the 8-inch gun Mk. VI Mod. 3A2, is a percussion-electric type firing mechanism. However, the gun should normally be fired electrically for safety reasons. The firing lock fits over the rear end of the obturator spindle. The primer is held in place, when inserted, by a wedge which slides up and down (to close and open) much like a wedge type of breech mechanism used with fixed ammunition. A firing pin extends through the firing lock wedge. The latter is raised or lowered by twisting the firing lock hammer. Extractors automatically withdraw the fired primer when the wedge is lowered.

b. To insert the primer, the firing lock hammer (fig. 52) is pulled back about 1/16 inch and twisted to the left. Then the hammer is twisted back to the right, bringing it into contact with the rear end of the firing pin (fig. 53). The firing lock is now ready for electrical firing. The current will pass through the hammer and firing pin to the primer.

c. To fire by percussion, the firing lock wedge is opened, the primer inserted, and the wedge closed as for electrical firing. Then the hammer is gently pulled to the rear and turned to the left about 45°. To compress the firing spring, the hammer is again pulled to the rear. The hammer is now turned back (right) 45°, engaging a sear and placing the hammer in line with the firing pin. The firing lock is now cocked (fig. 54). When the trigger is pulled by a lanyard, the sear is disengaged from the hammer, the firing spring expands, and the firing pin, struck by the hammer, strikes the percussion cap of the primer. When using percussion firing, extreme care must be taken to insure that the breech is completely closed before cocking the firing lock.

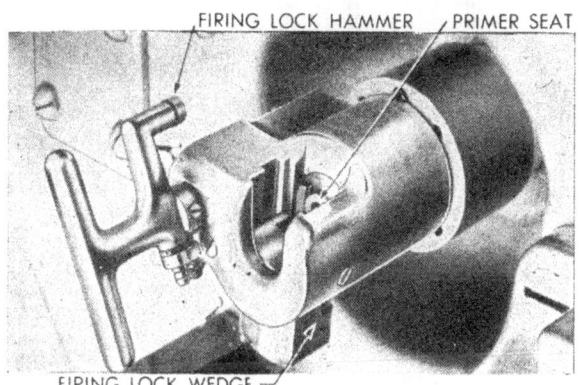

Figure 52. Firing lock Mk. VIII Mod. II (Navy type), open position, 8-inch gun Mk. VI Mod. 3A2.

d. As mentioned in a preceding, the gun should always be fired electrically when possible. There are no safety features provided for lanyard (percussion) firing. On the other hand, there are breaks in the electrical system which prevent electric firing until the breech is completely closed and the gun returned to battery.

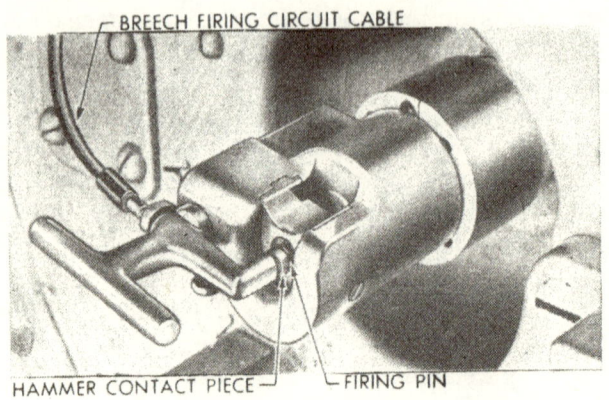

Figure 53. Firing lock Mk. VIII Mod. II, ready for electric firing, 8-inch gun Mk. VI Mod. 3A2.

Figure 54. Firing lock Mk. VIII Mod. II, cocked for percussion firing, 8-inch gun Mk. VI Mod. 3A2.

CHAPTER 3

RECOIL AND COUNTERRECOIL MECHANISMS

24. General

a. The recoil system is considered to be a part of the cannon carriage. However, for the convenience of discussion it is considered separately in this manual, but it will be referred to or considered briefly when discussing individual carriages in chapter 4.

b. On the early muzzle loaders the full force of recoil was taken directly by the carriage which consequently had to be very strong and heavy. When cannons were mounted on wheels, they were allowed to roll to the rear on each shot, absorbing some of the recoil by moving. The cannon were then returned to their battery position by the use of blocks and tackles. The Rodman gun (fig. 2) was provided with an upper and lower carriage. The upper was allowed to slide on the lower. As muzzle velocity increased and rapidity of fire became paramount, a new method of recoil braking was developed. The fundamental principle of the new method is the interposition of a recoil system between the cannon and the carriage. This brings the stresses on the carriage down to a reasonable value, insures stability, and permits the use of a lighter carriage (without rupturing or overturning). The carriage is designed so that the stress imposed by recoil is distributed over a convenient length of the recoil to prevent all of the shock from being transmitted to the carriage at the same instant.

c. Modern recoil brakes are designed to check the recoil of a cannon gradually and to finally stop it. To return the cannon to firing position, some form of automatic counterrecoil device is used in connection with the recoil brake.

d. As illustrated in figure 55, the modern recoil system has three distinct parts: the recoil brake (A), the counterrecoil mechanism or recuperator (B), and the buffer (C). The brake brings the cannon to a gradual stop in recoil; the counterrecoil mechanism returns the cannon to firing position (battery) and holds it there, regardless of elevation; and the buffer brings the cannon to a gradual stop when it returns to firing position.

25. Recoil Brakes

a. Although the friction of the cannon in the slide (par. 9) plus the energy absorbed by the counterrecoil system contribute a small amount of resistance to the recoiling cannon, the greater part of the force of recoil is received by the brake itself (par. 24*b*).

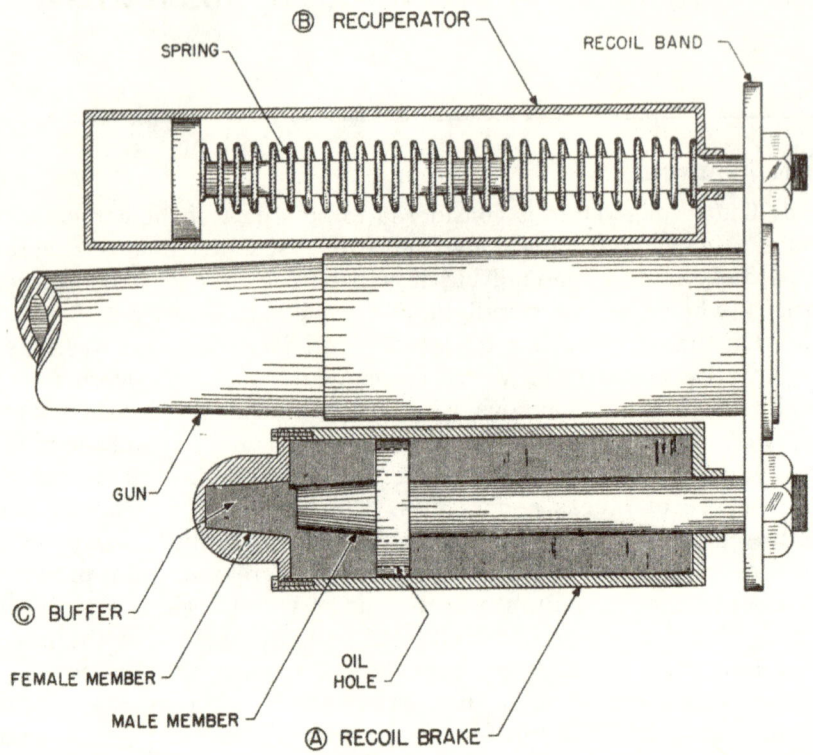

Figure 55. Simplified recoil system, schematic drawing.

b. The modern recoil brake consists basically of a piston in a cylinder filled with oil (A, fig. 55). On recoil, the piston rod, which is attached to the cannon, moves the piston to the rear forcing oil at high pressure through small holes in the piston or grooves in the cylinder wall. *The resulting resistance gradually brings the cannon to a stop.* In some systems the cylinder is attached to the cannon and moved with it during recoil, while the piston is fixed to the carriage and remains stationary. Relative motion between piston and cylinder is fundamentally the same in both systems.

c. Oil holes (apertures) of constant area do not provide uniform braking; recoil velocity is too low at the start and too high at the finish. This results in unbalancing the cannon and carriage. In order to obtain uniform recoil it is necessary to vary the size of the apertures so that their area is proportional at any point to the velocity of recoil at that point. In other words, the area of the apertures should be larger when the velocity of the piston is greater and diminish when the velocity of the piston diminishes. In designing the recoil system the variation in the area of the apertures is fixed so as to keep the total resistance to recoil constant, or to make it vary in any manner desired, throughout the length of recoil. There are several means by which the area of the apertures can be regulated so as to control the rate of flow of the oil and thus control the resistance offered to recoil. Several devices for throttling the oil flow are as follows:

(1) *Throttling grooves.* Grooves (fig. 56) of varying width or depth are cut in the wall of the cylinder so as to regulate the rate of flow to suit the velocity of recoil. The piston is without holes; hence, the oil is forced to pass through the throttling grooves under pressure of the motion of the piston. The grooves are slotted in such a manner as to have the greatest depth (or width) at the beginning

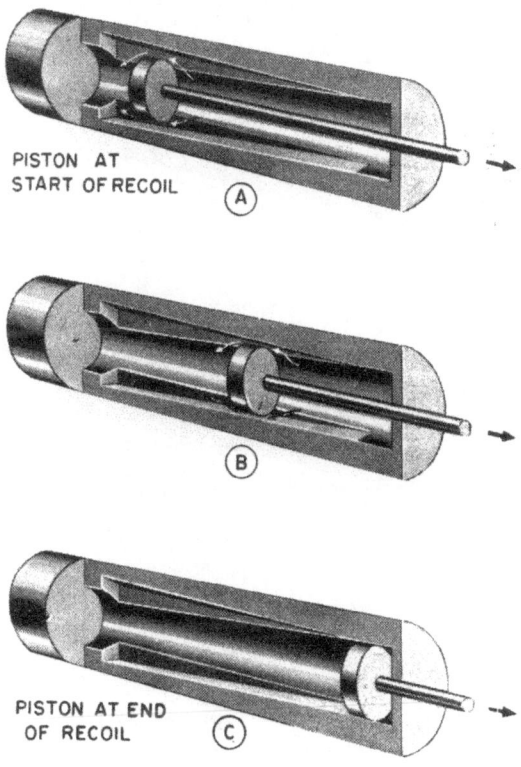

Figure 56. Throttling grooves.

of recoil and a decreasing depth along the grooves until there is none at all at the end of recoil. Therefore, as the piston is drawn by the force of recoil to the rear and the velocity decreases, there is a proportional decrease in the flow of oil, bringing the recoiling mass to a gradual stop. At the end of recoil, the solid piston completely seals the cylinder. The 16-inch howitzer and some 16-inch guns use this type of recoil brake. The number of grooves is usually three.

(2) *Throttling bars.* (a) Throttling bars (fig. 57) are wedge-shaped bars set in the walls of the cylinder and function in rectangular slots cut in the piston. The bars work in much the same manner as the throttling grooves. As the gun starts to recoil, the notches in the piston form apertures through which the oil can pass under pressure. The size of the apertures decreases as the cannon recoils since the bars are of variable height and constructed to gradually increase until the apertures are completely closed at the end of recoil, thus easing the cannon to rest. Throttling bars are found on some of the older models of seacoast guns.

(b) A variation of this device uses tapered throttling rods instead of bars. These are attached to the ends of the cylinder. The tapered rods run through round holes in the piston instead of rectangular

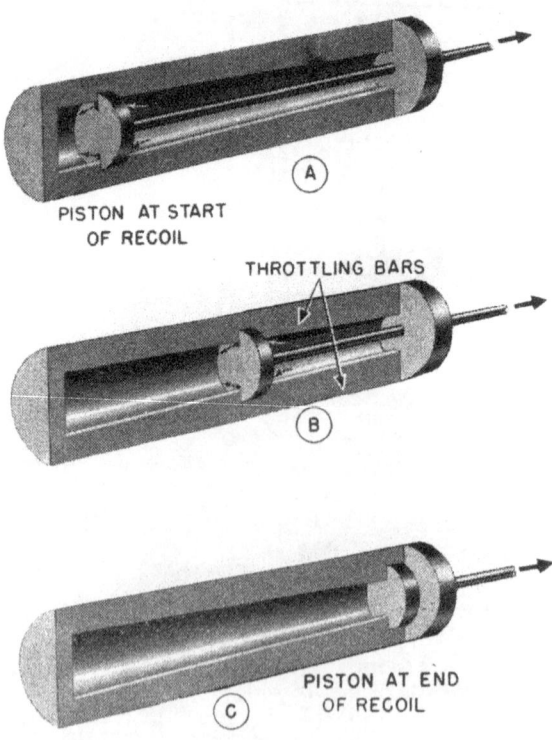

Figure 57. Throttling bars.

notches. The action is the same in both cases. Three rods (fig. 58) of this type are used in the recoil cylinder of the 16-inch gun Mk. II M1.

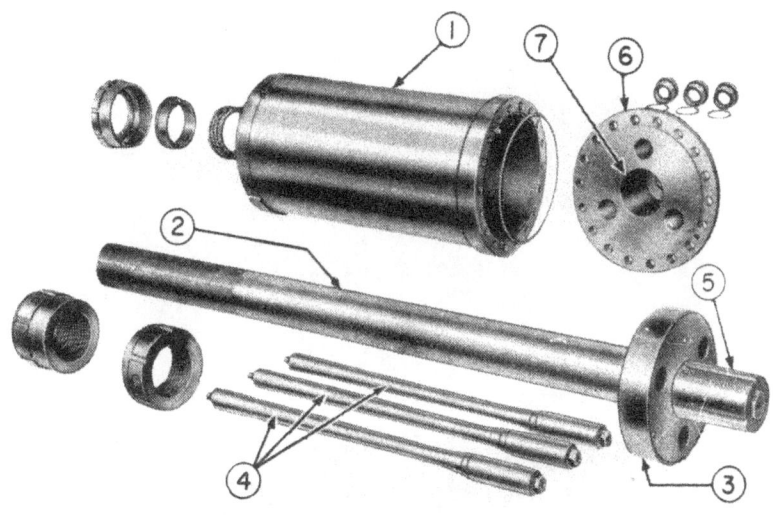

1. Recoil cylinder.
2. Piston rod.
3. Piston.
4. Throttling rods.
5. Buffer plunger.
6. Cylinder head.
7. Dashpot recess.

Figure 58. Recoil cylinder, 16-inch gun Mk. II M1, exploded view.

(3) *Throttling rod.* In this system a single tapered throttling rod (fig. 59) is attached to the end of the cylinder. This throttling rod is received by a hollow piston rod. Orifices are provided in the piston for the flow of oil. During recoil the piston and piston rod move to the rear of the cylinder. At the beginning of the movement, the smallest section (least circumference) of the throttling rod is opposite the orifices, thus providing space for the free flow of the oil in the initial stages of recoil. As the gun recoils, the rate of flow of the oil decreases because the clearance between piston apertures and the throttling rod is decreased.

26. Variable Recoil

a. GENERAL. Because of the increased ranges demanded in modern warfare, there is a need for increased cannon elevation. To provide for such a situation, recoil brakes must be able to return the cannon to battery at all elevations. At the same time (especially in mobile weapons, par. 32) it is necessary to control the length of recoil (fig. 60) in such a manner as to bring about long recoil in horizontal firing to assure stability, and short recoil at high-angle firing to prevent the cannon from striking the ground. Such is the purpose of the variable recoil mechanism. In discussing the variable recoil system, the recoil mechanism employed on the 155-mm gun M1 will be used as an example.

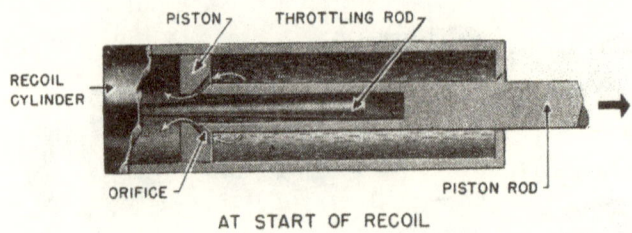

AT START OF RECOIL

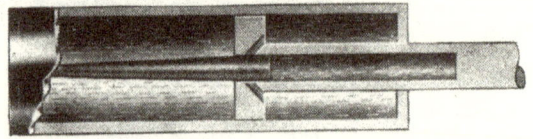

AT END OF RECOIL

Figure 59. Throttling rod.

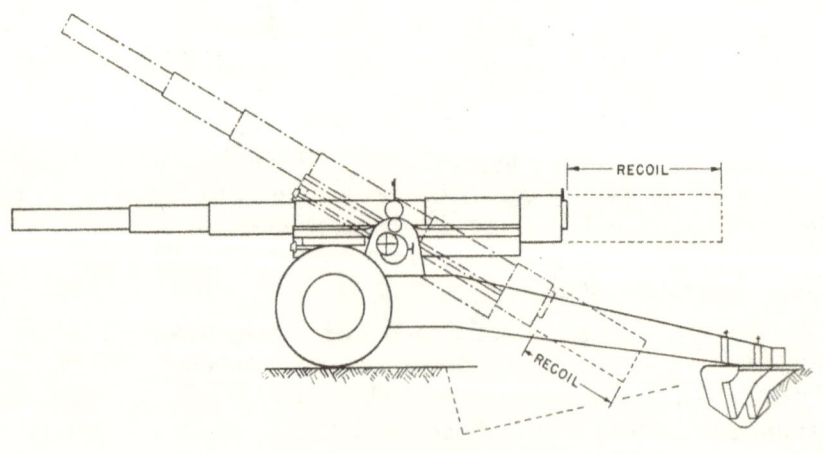

Figure 60. Variable recoil.

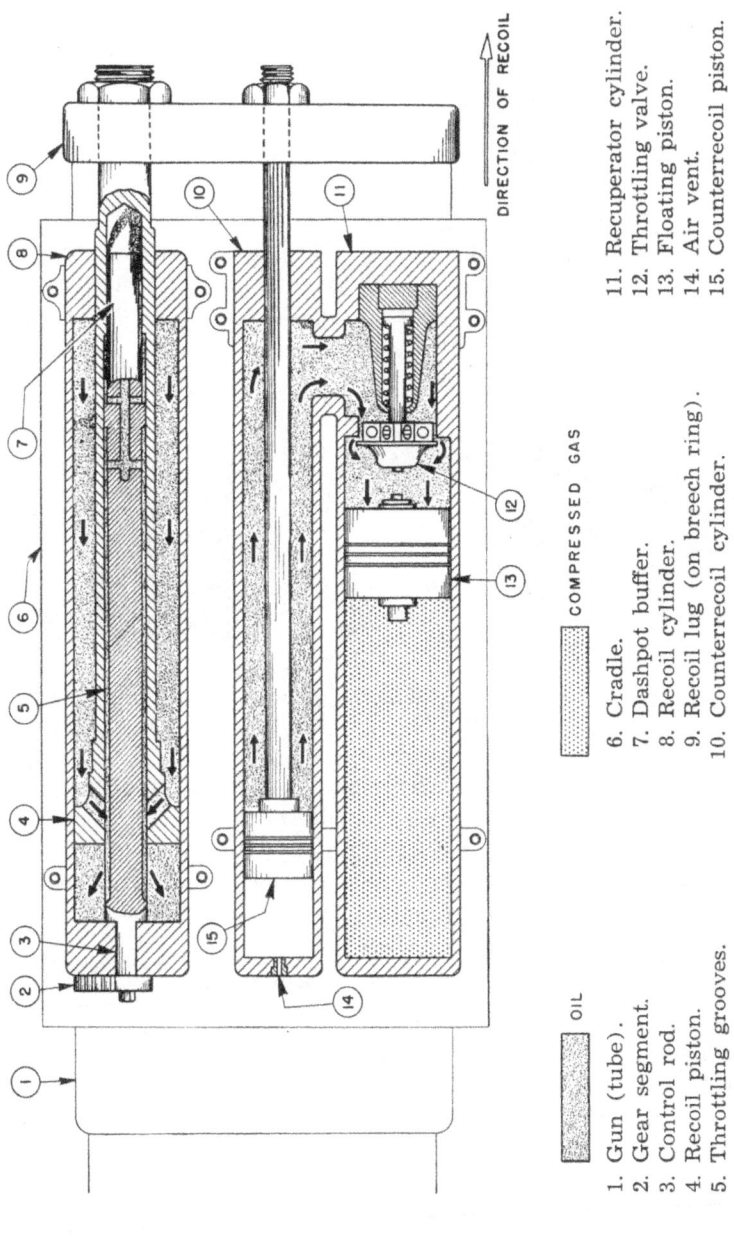

Figure 61. Recoil and counterrecoil mechanisms, 155-mm gun M1, schematic drawing showing flow of oil during recoil.

b. OPERATION. (1) The recoil cylinder (fig. 61) contains a piston on a hollow piston rod which is attached to the gun and moves with it in recoil. Ports in the piston provide paths for the flow of oil from the rear side of the piston to the forward side during recoil. A control rod, which can be rotated but not moved longitudinally, is located inside the piston rod and attached to the forward end of the recoil cylinder. Throttling grooves of varied depth cut in this control rod provide maximum flow of oil at the beginning of recoil and a gradual decrease to zero at the end. In addition, the grooves are so arranged that rotation of the control rod will also vary the area of the apertures. It is this rotation of the control rod that provides variable recoil—from approximately 33 inches at maximum elevation to 65 inches at zero elevation. That is to say, the length of recoil is automatically shortened as the angle of elevation of the gun is increased.

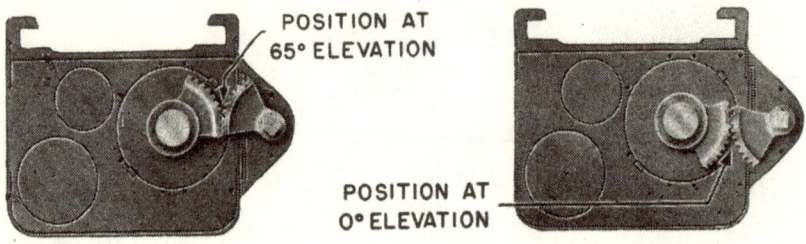

Figure 62. Variable recoil mechanism, front view of recoil cylinder, 155-mm gun M1.

(2) The rotation of the control rod is accomplished by the movement of a gear segment (fig. 62) on the forward end of the rod. This gear segment is engaged by a similar segment which is connected to the top carriage by the variable recoil linkage (figs. 63 and 144) in such manner that the control rod is positioned automatically by the elevation of the cradle.

c. REPLENISHER. A spring-actuated replenisher (fig. 63) is connected to the front end of the recoil cylinder of the 155-mm gun M1. It serves as a reservoir for excess oil when increased atmospheric temperature or heat, developed during firing, expands the oil in the recoil cylinder. It also keeps the recoil cylinder filled when the oil contracts because of decreased temperature, or when there are slight leaks around the packing. Furthermore, during recoil a void is produced in the recoil cylinder and the replenisher serves to fill it. When the gun returns to battery, the oil is again forced back to the replenisher.

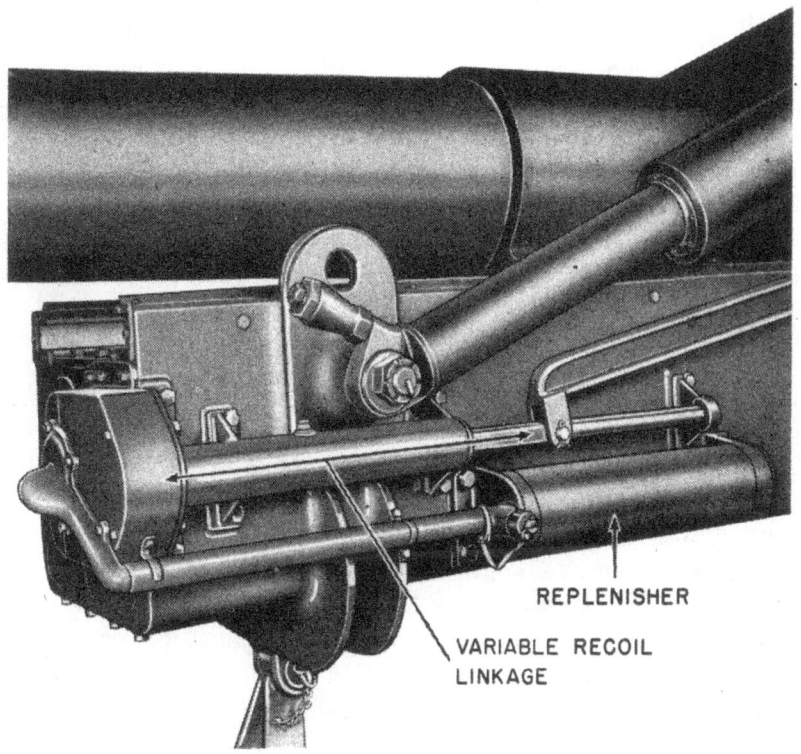

Figure 63. Variable recoil mechanism and replenisher, 155-mm gun M1.

27. Counterrecoil or Recuperator Systems

a. GENERAL. Recuperator or counterrecoil mechanisms are employed on cannon mounts to return the cannon to its firing position after the recoil mechanism has brought the recoiling mass to a halt. Energy for this purpose is secured from the momentum of the cannon during recoil. This energy is stored in a suitable medium such as springs or compressed air which also must be initially in a state of compression to prevent the cannon from slipping back in the cradle when elevated. In addition, counterrecoil action must be controlled to prevent it from returning the cannon to battery with progressive force capable of damaging the piece. The counterrecoil system also adds resistance to recoiling. However, such resistance is negligible as compared to the action of the recoil brake itself.

b. TYPES. The return of the cannon to battery after recoil may be effected by coil springs, gravity, or compressed gas.

(1) *Spring.* (*a*) Spring recuperators (B, fig. 55) consist essentially of spiral springs which are compressed within cylinders when the cannon recoils, thus storing up energy. The coil springs may be

arranged concentrically for greater power or telescopically for greater length of recoil, but the principle of operation does not change. These springs bear against the head of the cylinder at one end and against the piston at the other end. In recoil, the springs are compressed by the piston which is attached to the cannon by a piston rod. At the end of recoil the contracted springs expand to their normal shape, returning the cannon to battery.

(b) The merit of a spring recuperator is its simplicity, but its required over-all length prevents its use on long recoil carriages unless a telescopic system is used. The 12-inch gun M1895M1A4 on barbette carriage M1917 uses a spring system consisting of four cylinders each containing three concentric columns of helical springs. Each column is divided transversely into four sections. Each section is separated from the adjacent section by a spring separator, which is simply a diaphragm not attached to the piston rod or to the cylinder. Figure 113 shows the arrangement of the springs in the recuperator system of 6-inch gun barbette carriage M1.

(2) *Gravity type.* The gravity method, used on the obsolete disappearing and some barbette carriages, is being discarded.

(3) *Pneumatic recuperators.* (a) The pneumatic recuperator consists of a cylinder filled with gas (air or nitrogen) under high pressure and a piston operating within the cylinder. When the cannon recoils, the gas in the cylinder is still further compressed. This pressure, acting on the piston, returns the cannon to firing position.

(b) Gas under high pressure is difficult to imprison, and several means have been tried to obtain airtight packing. The most successful means has been the oil seal. This device makes use of the fact that it is comparatively easy to design a packing to confine oil but difficult to design one to confine gas. As shown in figure 61, an auxiliary or floating piston may be employed to separate the oil from the high-pressure gas. Another method has the oil in direct contact with the gas, but prolonged firing sometimes causes the gas and oil to form an emulsion (froth). Consequently, the method is not generally used. In the counterrecoil mechanism shown in figure 61, there are two cylinders entirely separate from the recoil cylinder (par. 29c). In some systems there is only one cylinder, also separate from the recoil cylinder. On the other hand, some counterrecoil mechanisms (fig. 67) are connected to the recoil brake. In this arrangement, the oil from the recoil cylinder flows into the recuperator cylinder when the gun is in recoil, compressing the gas within the recuperator cylinder. In counterrecoil the gas expands against a floating piston, forcing the recoil oil back into the recoil cylinder.

28. Counterrecoil Buffers

a. The force required to return a cannon from recoil to firing position is considerable; moreover, it must be capable of returning the cannon to firing position at its highest usable elevation. If the cannon is fired at 0° elevation, there is merely the friction of the slide to overcome, and counterrecoil velocity is much greater than when fired at high elevation. To absorb the excess energy of counterrecoil and bring the cannon to a gradual stop regardless of elevation, the counterrecoil buffer is used.

b. Most buffers are of the dashpot or plug type, as shown in C, figure 55. During the early part of counterrecoil, oil is trapped in the female part of the buffer. During the latter part of counterrecoil, this oil must escape through the narrow opening between the male and female members. As counterrecoil nears completion, the opening between the male and female members becomes increasingly smaller and finally disappears. The force expended in forcing the oil through the small opening gradually slows the cannon down and finally brings it to a stop. In addition to this "short control," some guns, such as 155-mm models, are provided with means for "long control" of the velocity of counterrecoil by throttling the return of the oil from the recuperator through special orifices. This applies a buffer action throughout counterrecoil. This mechanism is shown in figure 61. The throttling valve in the recuperator cylinder allows the oil to flow freely during recoil, but during counterrecoil it checks the flow of oil back into the counterrecoil cylinders. This keeps the compressed gas in the recuperator cylinder from expanding too rapidly.

29. Types of Recoil Systems

a. GENERAL. When any design of the hydraulic recoil brake is combined with any arrangement of counterrecoil springs, the recoil system is known as the hydrospring type. In like manner, when any type of the hydraulic recoil brake is combined with some form of the pneumatic recuperator, the recoil system is called the hydropneumatic type.

b. HYDROSPRING TYPE. An example of the hydrospring recoil mechanism is that used on the 3-inch seacoast gun carriage M1903 (fig. 64). This is a combination recoil, counterrecoil, and buffer system. There are two cylinders arranged concentrically with a counterrecoil spring coiled between the interior of the larger (spring cylinder) and the exterior of the smaller (recoil cylinder). The inner (recoil) cylinder is attached to the gun and moves to the rear when the gun is fired, compressing the spring. However, the greater part of the energy of recoil is absorbed by the resistance offered to the oil being forced through the throttling grooves in the wall of the recoil cylinder. The piston and piston rod, which contain the recess of the dashpot buffer system, remain stationary. The counterrecoil buffer is attached to the recoil cylinder and moves with it. As is seen in the figure, it is tapered so that the resistance to the escape of oil from the dashpot will gradually bring the gun to rest in battery. The length of recoil is about 9 inches.

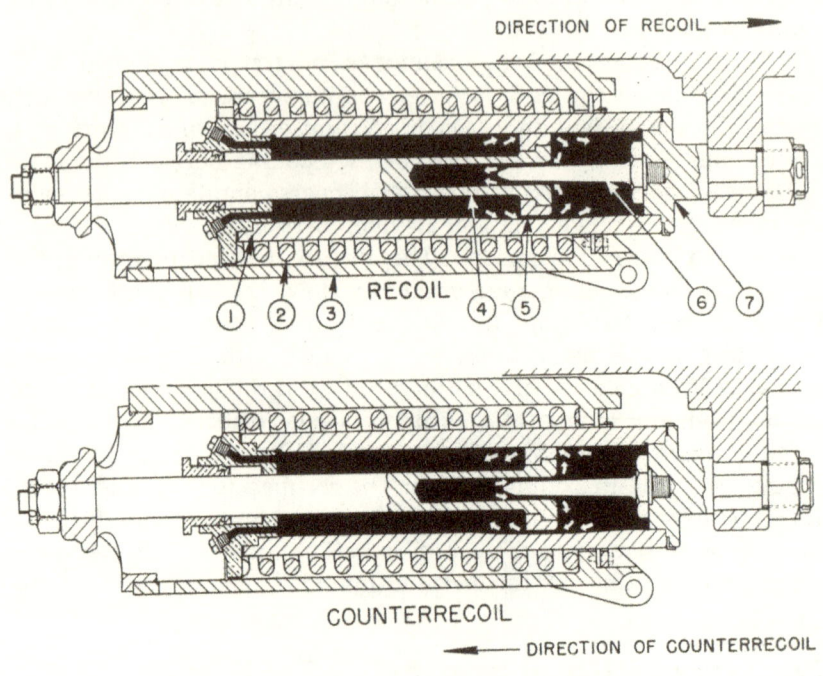

1. Recoil cylinder.
2. Counterrecoil spring.
3. Spring cylinder.
4. Piston rod and piston.
5. Throttling groove.
6. Counterrecoil buffer.
7. Rear recoil cylinder head.

Figure 64. Flow of oil during recoil and counterrecoil, recoil mechanism, 3-inch gun barbette carriage M1903.

c. HYDROPNEUMATIC TYPE. The 155-mm gun M1 recoil mechanism is of the hydropneumatic type and is composed essentially of the recoil cylinder, the counterrecoil cylinder, and the recuperator cylinder all assembled in the cradle (fig. 61). The action of the recoil brake is explained in paragraph 26; the counterrecoil and buffer system is partially explained in paragraphs 27b and 28b. Briefly, the action of the counterrecoil system is as follows: When the gun is fired, the oil in the counterrecoil cylinder is forced into the recuperator cylinder and pushes the floating piston forward to compress the nitrogen gas in the forward end of the recuperator cylinder. This action absorbs some of the energy of recoil and stores it for the counterrecoil movement. At the end of recoil, the pressure of the compressed gas forces the floating piston in the opposite direction (rearward) and the oil in the rear end of the recuperator is forced through the throttling valve back into the counterrecoil cylinder against its piston, thereby forcing the recoil piston and gun into firing position. The velocity of counterrecoil is controlled by the action of the throttling valve in the recuperator cylinder as explained in paragraph 28*b*. Besides this mechanism there is also a dashpot buffer system provided by the action of the rear end of the control rod in the hollow piston rod.

30. Recoil System, 16-inch Gun Mk. II M1

a. RECOIL BRAKE. The recoil brake on the 16-inch gun Mk. II M1 (fig. 58) is similar to the conventional hydraulic type as described in paragraph 25. There is one recoil cylinder (fig. 83) located on the under side of the cradle. Three stationary throttling rods extend through apertures in the piston. The diameter of each rod varies along its length; thus, the area of the clearance between the piston apertures and the throttling rods varies with the position of the piston during recoil. The resistance built up as the liquid (glycerin and water) is passing through the apertures, plus the resistance of the compressed air in the recuperator cylinders, produces a constant resistance throughout the length of recoil (normally 48 inches). An expansion chamber (not shown), attached to the cradle and connected to the recoil cylinder, acts as a reservoir to receive the overflow of liquid from the recoil cylinder due to heat expansion.

b. COUNTERRECOIL. (1) The recuperator mechanism consists of three independent cylinders (fig. 83), each a complete counterrecoil unit. Each cylinder (fig. 65) contains two gas (air) chambers located in the rear portion of the cylinder and separated by a check valve mechanism. In the forward end of the cylinder is a hollow plunger filled with a seal of glycerin and water. A floating piston rod extends through this plunger into the cylinder. On the end of the rod is a floating piston, separating the glycerin mixture from the air. The plunger is attached to the gun and recoils with it when the latter is fired. The initial air pressure in the counterrecoil cylinder is sufficient to hold the gun in battery at all elevations.

(2) When the gun fires and recoils, the plunger is forced back into the cylinder, causing the pressure in the air chambers to increase rapidly. The increased pressure offers resistance to the movement of the recoiling parts and builds up a sufficient force to return the gun to firing position at all angles of elevation. On recoil, the

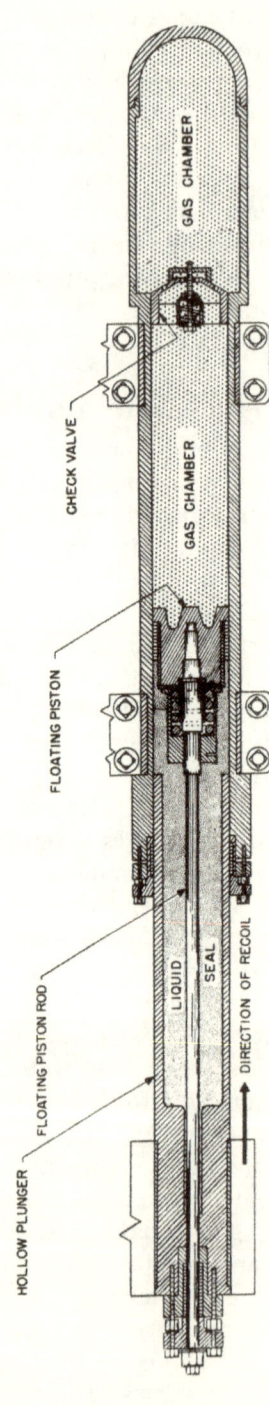

Figure 65. Counterrecoil cylinder, 16-inch gun Mk. II M1, barbette carriage M4, schematic drawing.

check valve opens automatically, allowing the air in the front chamber to pass freely into the rear chamber as it is compressed by the floating piston. However, when the gun begins the counterrecoil movement, the valve closes automatically and the compressed air in the rear chamber is allowed to pass through small openings in the valve. This action retards the flow of gas into the front chamber and thereby diminishes the velocity of counterrecoil. Long control is thus provided.

(3) Short control of counterrecoil is provided by the buffer action of a plunger attached to the front end of the recoil piston rod and piston (fig. 58). This buffer plunger enters the dashpot recess in the recoil cylinder head during counterrecoil when the gun is out of battery a distance of 12¾ inches.

31. Recoil System, 90-mm Gun M1

a. GENERAL. The recoil mechanism on the 90-mm gun M1 (fig. 66) is variable and of the hydropneumatic type. It is composed of the recoil cylinder, floating piston cylinder, and gas cylinder assembled in the lower part of the cradle, and a counterrecoil buffer mechanism attached to the top of the cradle.

b. OPERATION. (1) The recoil cylinder (fig. 67) contains a solid piston and piston rod attached to the breech ring and moves with the gun in recoil. The floating piston or recuperator cylinder contains recoil oil and nitrogen gas separated by a floating piston. The initial pressure of the gas is 830 pounds per square inch. The recoil and recuperator cylinders are connected so that the recoil oil can flow from one to the other. However, this flow of oil is controlled by two valves, the throttling valve for recoil control and the counterrecoil valve. The throttling valve is automatically controlled by the elevation of the gun. In firing, the recoil piston slides to the rear of the recoil cylinder. The oil in that cylinder is then forced by the rearward pressure of the recoil piston to pass into the floating piston cylinder through the orifice regulated in size by the throttling valve mechanism, while at the same time the counterrecoil valve is closed. The pressure of the oil on the floating piston causes the gas in the cylinder to be further compressed. At the end of recoil the throttling valve is closed and the counterrecoil valve is opened. The expansion of the gas forces the oil back into the recoil cylinder and returns the gun to firing position. The gas cylinder (fig. 66) is, in effect, an extension of the floating piston cylinder. It contains only gas under pressure and is connected to the floating piston cylinder by a by-pass connection.

(2) Variable length of recoil is accomplished by the action of a control bar assembly (not shown) on the recoil throttling valve

stop (fig. 67). The pressure of the control bar against the stop regulates the size of the orifice. The control bar is actuated by a cam (not shown) mounted on the inside of the top carriage. Normal recoil at 0° elevation is 40 to 46 inches; at 80° elevation, it is 24 to 26 inches.

c. COUNTERRECOIL BUFFER MECHANISM. This mechanism is separate from the recoil and recuperator cylinders and consists of an oil-filled cylinder with a spring, piston, piston extension, and packing at each end (figs. 66 and 68). When the gun is fired, the spring between the front of the cylinder and the piston forces the piston and piston extension to follow the breech ring in recoil. The valve at the piston moves back and uncovers oil passage holes in the piston, allowing the oil to flow freely to the spring side of the piston. In counterrecoil the breech ring drives the piston forward and causes the valve to cover the oil passage holes. The oil must now transfer back to the rear of the cylinder through grooves in the cylinder wall which regulate the flow of oil to slow down and stop the counterrecoil movement of the gun.

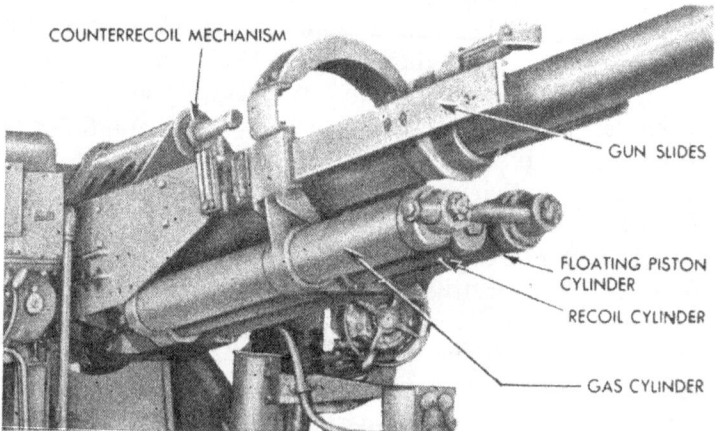

Figure 66. Gun slides and recoil mechanism, 90-mm gun M1.

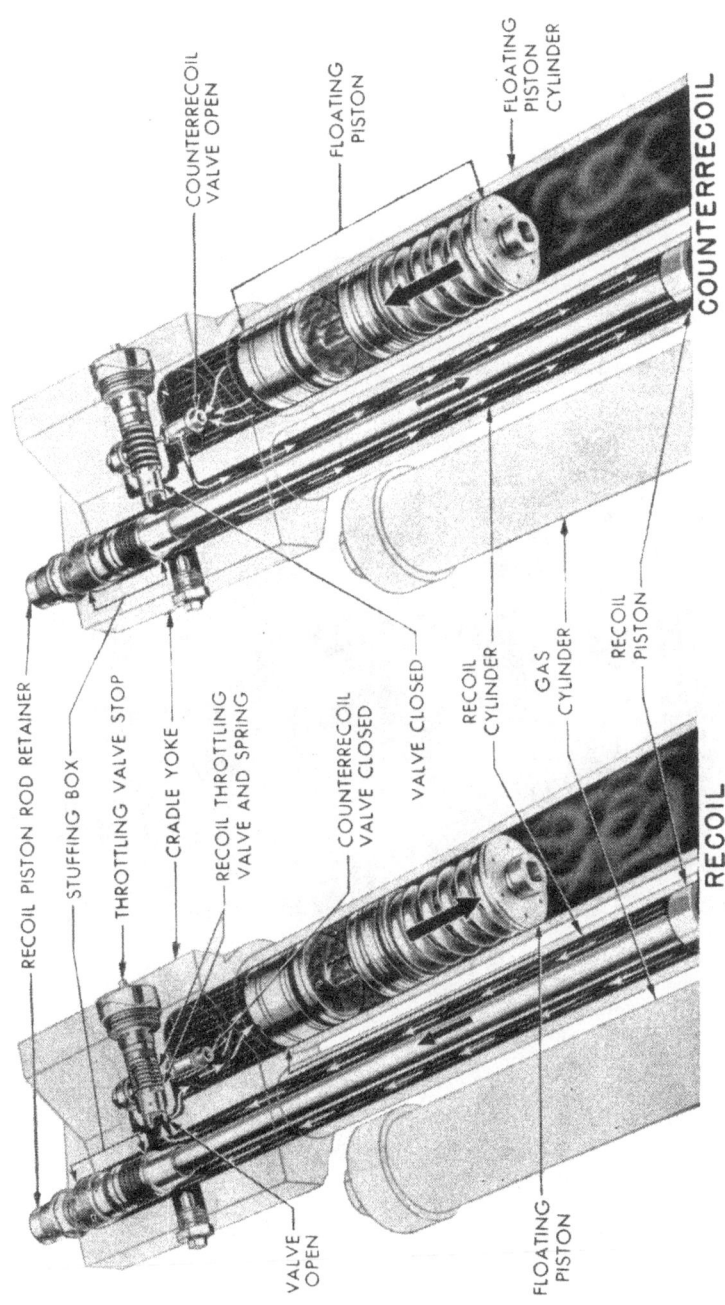

Figure 67. Flow of oil in recoil and counterrecoil, 90-mm gun M1, schematic drawing.

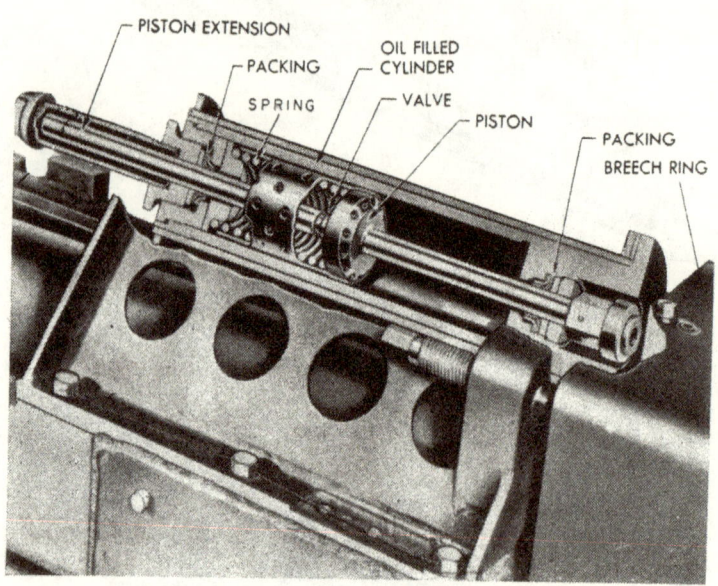

Figure 68. Counterrecoil buffer mechanism, 90-mm gun M1.

CHAPTER 4

CARRIAGES

Section I. GENERAL

32. General

Modern seacoast artillery carriages (mounts) may be divided into three types: barbette, tractor-drawn, and railway. The first type is fixed, the other two are mobile. Though differing somewhat in size, appearance, and method of employment, all of these types have many points of similarity. Nevertheless, there are certain fundamental differences such as the weight and the length of recoil. In a mobile mount of the tractor-drawn type, weight is an important factor. The lighter weapon can negotiate weaker highway bridges, is less likely to get stuck during wet weather, and is in general a handier piece of artillery. To be able to cut down on the weight of the carriage, it is necessary to cut down on the stresses that the carriage is expected to bear, and one of the easiest ways to do this is to lengthen the recoil. Therefore, our mobile cannon which must be pulled along the road are all long-recoil weapons. If it is desired to use cannon at high angles of elevation, a compromise must be reached between the advantages of long recoil and the disadvantages of having to dig a deep pit into which the cannon may recoil without hitting the ground. This is worked out in the 155-mm guns by having variable recoil—long (approximately 6 feet) at low angles of elevation and short (approximately 3½ feet) at maximum elevation. At the other extreme are the fixed barbette mounts bolted to concrete blocks. Here stability and weight take care of themselves, and a short, constant length of recoil may be adopted. For example, the length of recoil on the 6-inch gun, barbette carriage M1, is a little over 1½ feet. A study of the length of recoil of artillery weapons in various armies shows that mounts of the same general caliber and function have very similar lengths of recoil. The railway carriage is between the fixed and the tractor-drawn mounts in this respect, and generally employs a constant length of recoil longer than that of the fixed mount.

Section II. FIXED CARRIAGES

33. General

For a more adequate understanding of the construction of fixed seacoast cannon, a familiarity with some of their basic advantages and disadvantages is essential. Usually, fixed cannon are located in important strategic areas and in the best available tactical positions. They employ the use of concrete and earth emplacements as a protection for personnel and equipment. They can attain a high rate of accurate fire, since the base-end stations, observation stations, plotting rooms, powder rooms, and ammunition systems are permanently established. Their disadvantages lie in their inability to be moved to another position in case of necessity and in the fact that the enemy is likely to know their locations.

34. Emplacements

The protection afforded an artillery piece is of paramount importance. This has been manifested in the construction of forts throughout the history of warfare. Various methods have been used in the past. One of these was the disappearing carriage (fig. 69), now obsolete because of its limited capabilities. The disappearing carriage has a limited angle of elevation (20° max.) with a consequent limited range far below that of comparable guns on barbette carriages. The emplacement, designed to afford protection against flat-trajectory naval fire, is now vulnerable to attack by bombardment aviation and to fire from long-range naval guns whose shells have a high angle of fall. At the present time there are three solutions to the problem of protecting barbette-mounted guns:

a. OPEN EMPLACEMENT. The first solution is based on dispersion of the elements of the battery. In this scheme, the cannon are in the open (fig. 70) and widely separated. The magazines, power plant, and plotting rooms are dispersed and may be bombproofed. This system provides for a 360° traverse and firing at high elevations, but it offers no protection against direct hits and strafing. However, with adequate camouflage its value is considerable, also, shields can be added to the cannon which provide protection for key personnel. In the past, the Army has built barbette mounts in the open because of the greater cost of alternate methods.

b. CASEMATE. The second solution, the casemate (fig. 71), is a modification of the first. The structure permits elevating the gun for maximum range and facilitates bombproofing the power plant, ammunition storage and service, and plotting room. Since it is constructed of concrete, steel, and earth, it provides a defense against aerial and gunfire attack but limits the effective field of fire to less

Figure 69. Disappearing carriage emplacement, now obsolete.

Figure 70. Open emplacement, 16-inch gun barbette carriage M1919.

Figure 71   Casemate emplacement, 16-inch gun Mk. II M1, barbette carriage M4.

than 180°. The use of shields on the guns and closure plates extending from the sides of the front opening of the emplacement greatly increases the protection afforded against attack. Many of our open emplacements have been casemated.

   c. TURRET EMPLACEMENT. The third solution is the turret type of structure which is employed on heavy warships. This type of emplacement offers numerous advantages. The power plant and ammunition storage and service are all underground; a maximum use is made of labor-saving devices, greatly increasing rate of fire and cutting down on the number of men needed; the installation is easily gasproofed; and the crew members are all behind armor plate. Installations can be made which permit a 360° field of fire and use of elevations necessary for maximum range. Installations of this type have been made where the space available was limited (fig. 72) and they are particularly desirable for important locations where all-around fire against any seaward or landward attack is required.

Figure 72. Turret emplacement.

35. Barbette Carriage, General Characteristics

a. GENERAL. (1) According to a strict definition, a barbette carriage is a fixed carriage on which a cannon is mounted to fire over a parapet (fig. 73). However, at the present time the term "barbette carriage" is used in a broader sense to refer to a fixed carriage (regardless of whether or not the cannon fires over a parapet) which is capable of traversing through large angles except as limited by a protecting turret or casemate (figs. 70, 71, and 72). It may also be considered as the support of the cannon, consisting of a combination of several or all of the following major components: cradle and recoil system, top carriage (upper movable part), bottom carriage (lower fixed part), elevating mechanism, traversing mechanism, and loading mechanism (fig. 74).

(2) The advantages of such a carriage are:

(*a*) All-around fire—except as limited by emplacement.

(*b*) Elevations up to 65°.

(*c*) High-speed operation.

(*d*) Simplicity and ruggedness.

(3) In all modern barbette installations the cannon is mounted in the cradle. Recoil and counterrecoil (recuperator) systems are mounted parallel to the cannon, so that recoil takes place parallel to the axis of the bore regardless of the firing elevation.

Figure 73. 12-inch gun barbette carriage M1892, now obsolete.

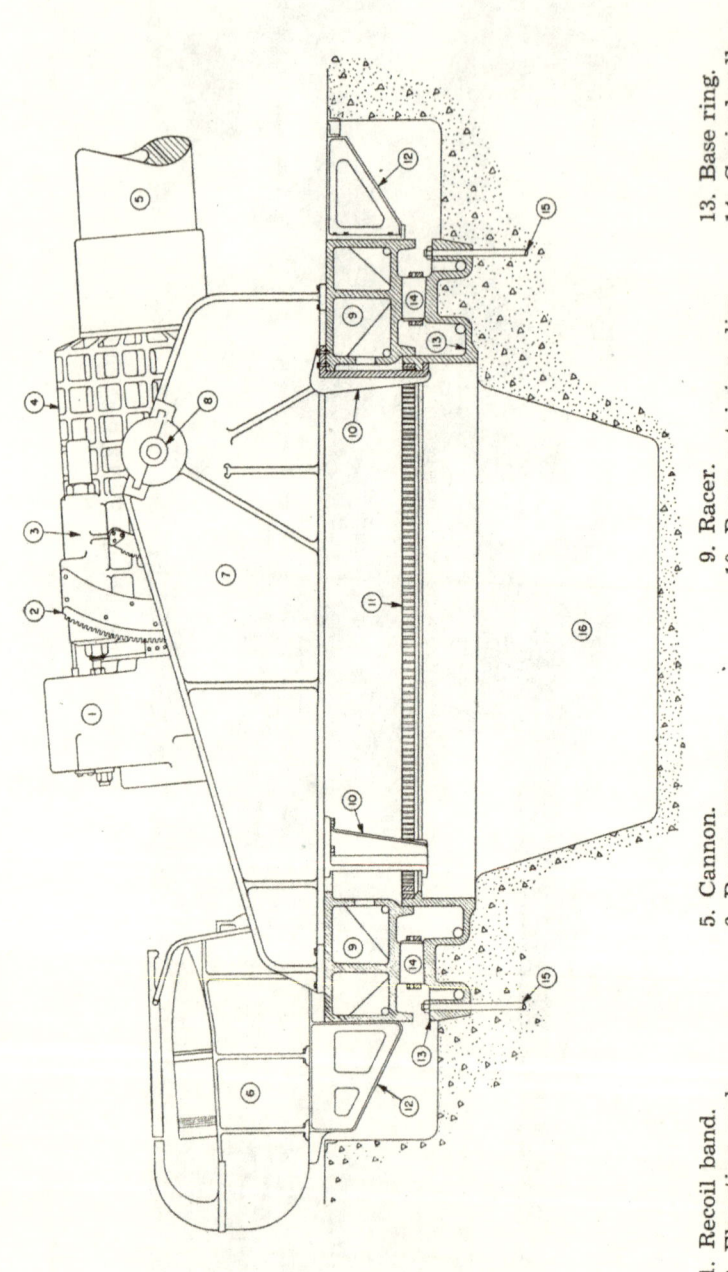

1. Recoil band.
2. Elevating rack.
3. Recoil cylinder.
4. Cradle.
5. Cannon.
6. Power rammer.
7. Side frame.
8. Cradle trunnion.
9. Racer.
10. Racer retaining clips.
11. Traversing rack.
12. Platform brackets.
13. Base ring.
14. Conical rollers.
15. Foundation bolts.
16. Recoil pit.

Figure 74. Barbette carriage.

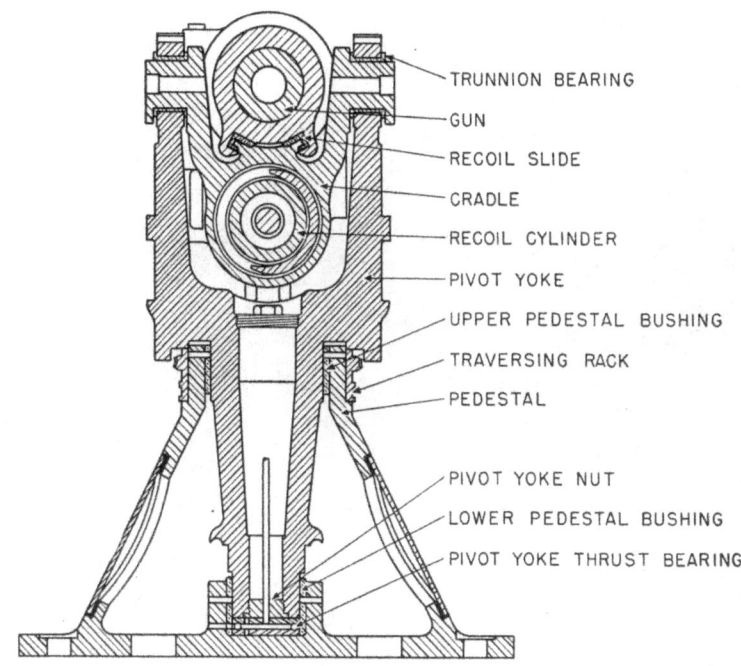

Figure 75. 3-inch gun barbette carriage M1903, pedestal type, sectional schematic drawing.

b. PEDESTAL TYPE. One type of barbette carriage used with smaller caliber guns is called the pedestal type. The general characteristics of this type of mount are shown in figures 75 and 76. A conical pedestal is bolted to the concrete platform. A pivot yoke, free to revolve, is seated in the pedestal. The upward extending arms of the pivot yoke form seats for the trunnions of the cradle. The cradle supports the gun, which slides on the cradle in recoil. The weight of all the revolving parts is supported by roller bearings on a central base within the pedestal. The recoil and recuperator cylinder (or cylinders) is (are) located in the lower rear portion of the cradle. A conventional recoil brake, spring recuperators, and a dashpot counterrecoil buffer are usually used on this carriage. On the 6-inch gun carriage, the brackets (to which are attached gunners' platforms) which move with the gun in traverse are bolted to the arms of the pivot yoke on each side.

c. CHARACTERISTICS OF MOUNTS, LARGER CALIBER CANNON. In general, the mount (fig. 74) for larger caliber cannon (as well as those for the 6-inch guns M1903A2 and M1905A2, barbette carriage M1) consists of a heavy base ring (fig. 80) bolted to a concrete emplacement and an upper carriage supporting the cannon and resting on the base. The upper carriage is capable of being moved in azimuth

Figure 76.—6-inch gun barbette carriage M1900, pedestal type.

upon the base. That is, the top of the base ring forms a path upon which are mounted conical rollers which support the superstructure and all traversing parts of the carriage. Resting on the rollers and revolving thereon is a racer (fig. 81) to which is bolted the upper carriage. To prevent the top carriage from tipping when the cannon is fired, racer retaining clips (fig. 74) are provided to hold the racer down to the base ring. These clips, bolted to the racer, form a band which is lipped inward at the bottom to traverse in the base ring groove. The trunnions (fig. 86) of the cradle rest in trunnion bearings in the upper carriage, permitting movement of the cannon and cradle in elevation.

  d. TRAVERSING AND ELEVATING MECHANISMS. (1) *General.* All carriages employ elevating and traversing mechanisms in order that they may be accurately set in elevation and direction. In most carriages of the same general design, these mechanisms have been standardized.

  (2) *Traversing mechanisms.* (a) Traversing mechanisms for major caliber weapons have, in most cases, become standardized on the type shown in figure 74. This mechanism is, in effect, a gigantic roller bearing with conical rollers operating between two bearing surfaces (lower, called the base ring; upper, called the racer). To the base ring, and concentric with it, is mounted a circular traversing rack (figs. 74, 80, and 108). A spur pinion, meshing with this rack, traverses the cannon. The efficiency of this system is illustrated by the 16-inch gun mount which requires a force of only 27 pounds at the traversing handwheel to traverse a mass of 660,000 pounds. An azimuth circle, also mounted concentrically with the racer, is provided for setting azimuth when firing at a target that cannot be

seen from the cannon position. For the larger caliber carriages, traversing is usually accomplished by electric power operating through Waterbury hydraulic speed gears (par. 36).

(b) To traverse minor caliber armament mounted on the pedestal-type carriage, some variation of a pivot yoke (figs. 75 and 119) revolving on antifriction bearings is used. To the upper part of the pedestal is bolted the traversing rack. A worm gear engaging this rack traverses the gun.

(3) *Elevating mechanisms.* There are two general types of elevating mechanisms: the elevating-rack type (fig. 117) and the screw-type (fig. 118).

(a) The elevating-rack mechanism (fig. 84) of latest design consists of a circular rack fastened to the cradle and a set of plain spur gears connected to the elevating handwheel on the main carriage. With a mechanism of efficient design it is possible to get a 1° change in elevation for each turn of the elevating handwheel and to elevate or depress a 16-inch gun with a force of 30 pounds at the wheel's rim.

(b) The screw-type elevating mechanism can best be illustrated by an ordinary screw automobile jack which transforms rotary motion at the handle into upward motion at the jack head. The jack base corresponds to the main carriage and the jack head to the cradle.

(c) Some form of protective device, such as a brake or overload slip device, is provided on all elevating mechanisms to prevent their injury from the shock of discharge and sudden stops and starts. On the barbette carriage this takes the form of a friction clutch which slips before the load on the gears becomes too great.

(d) Barbette carriages for major caliber cannon and 6-inch gun barbette carriages M1, M2, M3, and M4 are provided with electrical elevating mechanisms, operating through Waterbury hydraulic speed gears or Atlantic elevator equipment. Such equipment greatly decreases the time consumed in going from loading to firing position and vice versa.

(e) *Antifriction elevating device.* To enable major caliber cannon to be elevated and depressed quickly and easily, an antifriction elevating device (figs. 77 and 86) is employed. This is really a two-piece bearing—one, a roller bearing for easy operation; the other, a larger diameter main trunnion bearing to take the shock of discharge. During normal operation the main trunnion does not touch its plain main trunnion bearing at all, as the weight of the cannon is floated on a crutch and the roller bearing carries the load. When the cannon is fired, the force of recoil depresses Belleville springs so that the main trunnion rests on the main trunnion bearing, transmitting the force of recoil directly to the side frames of the carriage.

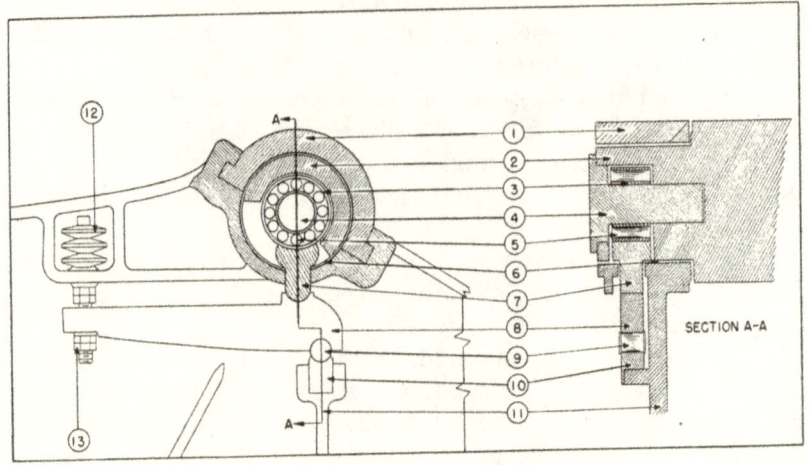

1. Main trunnion bearing.
2. Main trunnion.
3. Bearing sleeve.
4. False trunnion.
5. Roller bearings.
6. Clearance.
7. Crutch.
8. Lever arm.
9. Fulcrum pin.
10. Fulcrum seat.
11. Carriage side frame.
12. Belleville springs.
13. Adjusting nuts.

Figure 77. Antifriction elevating device.

When the cannon has returned to firing position, the Belleville springs again float the cannon's weight clear of the main bearing, and the roller bearing allows the cannon to be depressed or elevated easily.

e. POWER RAMMER. Power loading is used on major caliber weapons to increase the rapidity of fire and to insure uniform ramming which promotes greater uniformity in developed muzzle velocities. The rammer (fig. 88) consists of a steel frame, on top of which is a rammer tray or loading trough, and a flexible nonbuckling steel chain which is actuated by a motor through the medium of a hydraulic speed gear. If the motor should fail, hand power is supplied through two cranks located on the right and left sides of the rammer near the end. When the rammer is run forward, an unstroking device prevents the rammer head from advancing beyond a predetermined distance and returns the control lever to the neutral position. Likewise, on the withdrawal of the rammer an unstroking device performs the same functions. A spring buffer on the rammer head prevents excessive shock from injuring the mechanism. In operation, the movement of the rammer is controlled by means of a control lever. Raising or lowering the control lever from the neutral position puts the speed gear into operation.

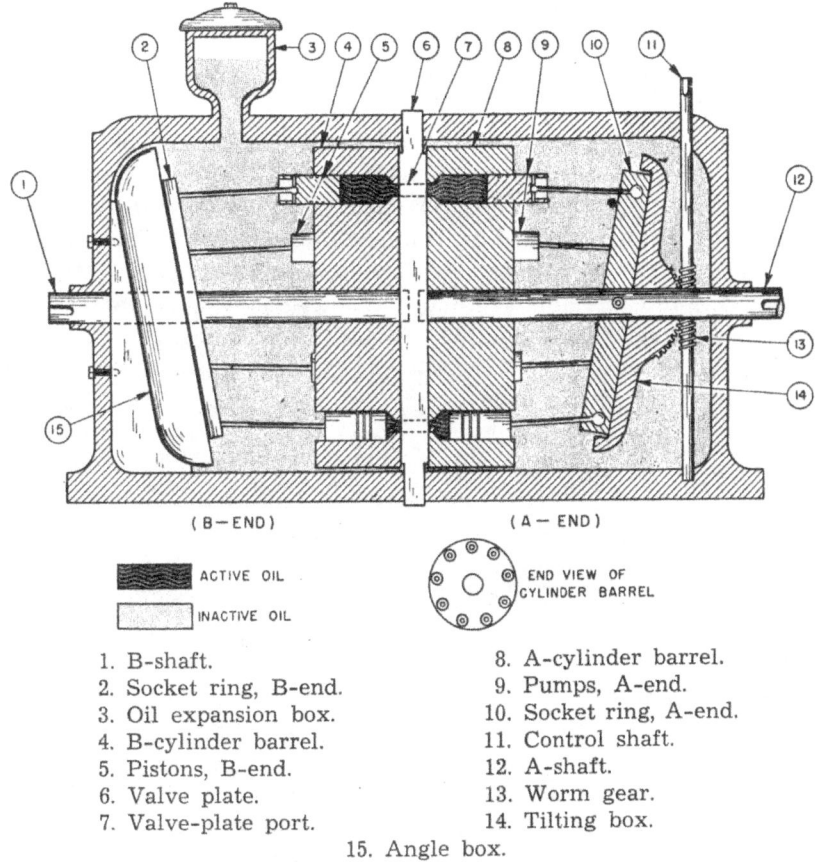

1. B-shaft.
2. Socket ring, B-end.
3. Oil expansion box.
4. B-cylinder barrel.
5. Pistons, B-end.
6. Valve plate.
7. Valve-plate port.
8. A-cylinder barrel.
9. Pumps, A-end.
10. Socket ring, A-end.
11. Control shaft.
12. A-shaft.
13. Worm gear.
14. Tilting box.
15. Angle box.

Figure 78. Waterbury hydraulic speer gear, simplified schematic drawing.

36. Elevating and Traversing Speed Control

a. WATERBURY HYDRAULIC SPEED GEAR. (1) Because electric motors for traversing, elevating, and ramming encounter maximum torque on starting, and because very fine variations in speed are necessary, conventional speed control is not satisfactory. Instead, the electric motor is allowed to run at its most efficient speed and is connected to the mechanism it is to operate through a Waterbury hydraulic speed gear. With this gear, any desired speed (either forward or backward) may be achieved while the driving motor runs continually at its designed speed in one direction.

(2) An illustration of the elevating mechanism and hydraulic system for 6-inch guns M1903A2 and M1905A2, barbette carriage M1, is shown in figure 109. A schematic sketch of the Waterbury hydraulic speed gear appears in figure 78. The right side (A-end) of the case is the driving side (a variable-delivery pump rotated by an

electric motor at a constant speed); the left side (B-end) is the driven side (a fixed-stroke hydraulic motor). The nine small cylinders, with plungers, of the A-end are small pumps arranged in a circular manner in the cylinder barrel. Oil fills all space within the case and valve plate that is not occupied by metal. A definite portion of the oil is enclosed within the cylinders ahead of the pistons and also within the port passages in the valve plate. It is this active oil, under pressure, which transmits the energy; the remaining or inactive oil is never under pressure and only serves as a supply for lubrication and replenishment. The tilting box, trunnioned in the case, does not rotate with the A-shaft but its angle of tilt may be changed as necessary to increase or decrease the amount of oil pumped through to the B-end. The A-socket ring is fixed to the A-shaft by a universal joint and rotates with it. The tilting box forms a guide or bearing for the A-socket ring. The socket ring is connected to the cylinders by connecting rods with ball joints at each end. The angle box in the B-end replaces the tilting box in the A-end and is always fixed in position at an angle of about 70°. Otherwise, the B-end is the same as the A-end. The two sides are connected by two valve-plate ports or by hydraulic piping as in the type shown in figure 109.

(3) As the A-shaft rotates the cylinder barrel and A-socket ring, more or less oil is pumped by the pumps, depending on the angle at which the tilting box is set by the worm gear. As the A-cylinders are moving down on the near side of the observer (fig. 78), oil is forced through the valve-plate port on this side into the B-cylinders of the near side. However, these cylinders cannot receive the oil unless their pistons move back. Thus, the backward movement is communicated to the inclined socket ring of the B-end through the reciprocating connecting rods, causing the B-shaft to rotate in a direction opposite to the rotation of the A-shaft. At the same time, the cylinders on the far side will draw oil through the port on the far side of the valve plate. The greater the tilt of the tilting box, the greater will be the stroke and the amount of oil pumped through to the driven side, and the faster the driven side shaft will rotate. As can be seen from the drawing, if the tilting box is tilted in the opposite direction, the driven shaft will rotate in the opposite direction; also, if the tilting box is perpendicular to the shaft, the pistons will not move with respect to their cylinders, there will be no oil pumped, and the driven side will remain stationary.

(4) By means of these gears a cannon may be elevated rapidly to its approximate elevation and then gently eased into its exact firing position; it may be depressed rapidly almost to horizontal and easily brought up against its stop for loading. An automatic stop is

Figure 79. 16-inch seacoast gun Mk. II M1, right side view.

provided to prevent injury to the gun due to careless depressing. Similarly, the mount may be traversed at will and the rammer operated with differing speeds for ramming projectiles and powder charges.

b. REMOTE CONTROL SYSTEM M14. (1) *General.* The remote control system M14 (Atlantic elevator equipment) is used on M2 and M4 6-inch gun carriages to supply controlled electric power for elevating the piece. This system performs the same function as the Waterbury hydraulic speed gear inasmuch as it eliminates torque in elevating and provides very fine regulations of elevating speed. In addition, the system may be operated so that it automatically sets the firing elevation as it is received over the data transmission system. Smooth speed control over a wide range is obtained by the use of a constant speed motor to drive a variable-voltage direct-current generator whose output is delivered to an elevating motor drive. In brief, the complete installation consists of the following major components:

(a) *A motor-generator set* which consists of a 3-phase, 440-volt alternating-current induction motor rated at 10 horsepower, a variable-voltage direct-current generator, and an exciter for supplying the generator and motor field currents. The alternating-current induction motor drives the generator and exciter.

(b) *A motor drive,* which is a 10-horsepower, variable-voltage, direct-current motor used to position the piece.

(c) *A control system,* the indicator-regulator M2, which converts the position of the selsyn receivers on the mount into suitable generator voltage for positioning the gun.

(2) *Operation.* (a) The elevating mechanism, when under automatic or semiautomatic control, is driven by the driving motor. The motor armature is supplied with direct-current voltage from the generator. This voltage may be varied both in amount and direction by the indicator-regulator. As a result, the motor armature turns in either direction at a speed proportional to the voltage supplied to it. The control elements of the indicator-regulator are positioned either by the data receivers or by the elevating handwheels. When the data receiver positions the control elements, operation is entirely automatic; whereas, when the handwheels position the control elements, the operation is semiautomatic (requires the elevation setter to operate the handwheels).

(b) A special feature of the Atlantic elevator equipment is the provision for depressing the gun to the predetermined loading position by throwing a single switch. When the loading operation is completed, a throw of the switch in the other direction results in the gun being positioned quickly (when elevating equipment is operating automatically) at any elevation established by the elevation receiver. When operating semiautomatically, the elevation setter must return the piece to the firing elevation by rotating the handwheels to position the control elements.

37. Barbette Carriage M4

a. GENERAL. The barbette carriage M4, with its 16-inch gun Mk. II M1, manifests practically all of the outstanding features of our modern major caliber guns (figs. 71 and 79). This carriage is a modification of barbette carriage M1919 (fig. 70). Previous modifications are designated M1919M1, M2, and M3. The latest modification is known as the M5. The carriage is provided with a 4-inch cast shield.

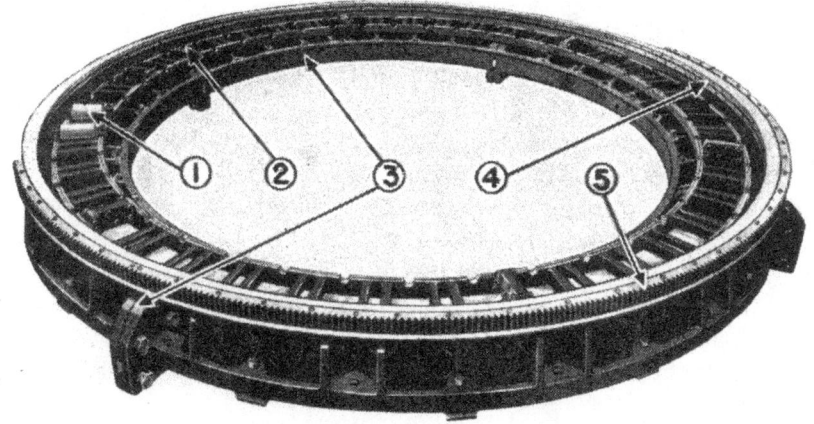

1. Traversing roller.
2. Traversing roller distance ring.
3. Base ring.
4. Base ring pintle liner.
5. Traversing rack.

Figure 80. Base ring and distance ring (two traversing rollers in place), 16-inch gun barbette carriage M4.

b. BASE RING. The base ring is composed of four sections bolted and keyed together and anchored to a concrete foundation by bolts on both the inner and outer flanges of the ring (fig. 80). A traversing rack is fastened to the outer annular flange of the base ring. An azimuth circle (fig. 82) is attached to the outer annular flange just below the traversing rack. A bronze liner, which forms the inner pintle surface of the base ring, is attached to the upper, inner, vertical section of the ring and furnishes a bearing between the base ring and the racer during the action of traversing (see *c* following). The upper surface of the base ring frame provides a path for the conical traversing rollers (held in place by the distance ring) which support the superstructure and all the traversing parts of the carriage.

c. RACER. Resting and revolving on the rollers is a racer which, like the base ring, is made of four sections bolted and keyed together (fig. 81). A vertical annular flange extends below the roller path and forms the inner pintle surface. This surface fits inside the

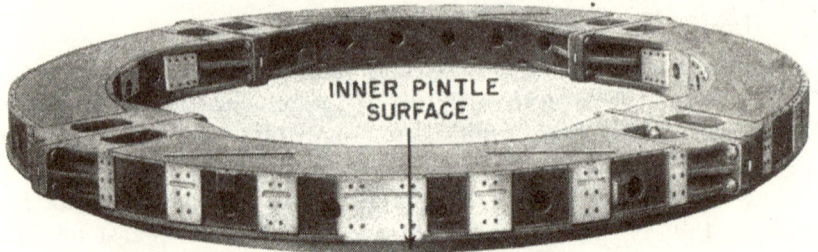

Figure 81. Racer, 16-inch gun barbette carriage M4.

bronze-lined base ring flange, forming the pintle for the carriage. Platform brackets (fig. 82), bolted to the outside surface of the racer, support the circular platform which surrounds the mount. Six of the brackets are constructed to serve also as racer clips by hooking under a projecting ledge on the base ring. This prevents any tendency of the racer to lift off the rollers when the gun is fired or when the gun is returning to battery in counterrecoil.

d. SIDE FRAMES. The side frames (fig. 82) provide trunnion bearings for the cradle and support for the tipping parts. Since they are rigidly bolted to the racer, they also provide support for the platform of the mount.

e. CRADLE AND THE RECOIL MECHANISM GROUP. The cradle and the recoil system were originally designed for the Navy and are lighter in construction than the Army cradle used on barbette carriage M1919. The cradle houses three recuperator cylinders located on the top and a single recoil cylinder attached to the under side of the cradle (fig. 83). The recoil and counterrecoil systems are described in paragraph 30.

f. ELEVATING MECHANISM. (1) *Hand and power elevation.* Two elevating racks (fig. 84) are provided on the cradle. The gun may be elevated by electric power or by hand. Elevating by electric power is accomplished by a motor acting through a hydraulic speed gear. Elevating and depressing cams, attached to the right rack, automatically stop the speed gear to prevent the gun from being elevated or depressed to the extreme limit. The motion of elevation or depression is controlled by the operator at the follow-up control handwheel. Elevating by hand merely requires slipping the clutch lever to the HAND position and thus engaging the gears of the hand-elevating mechanism. This mechanism is now in a position to transmit power through the same gear train as that used by the electric power system. The elevating handwheel on the right side of the carriage provides slow motion; an elevating crank on the left side provides fast motion.

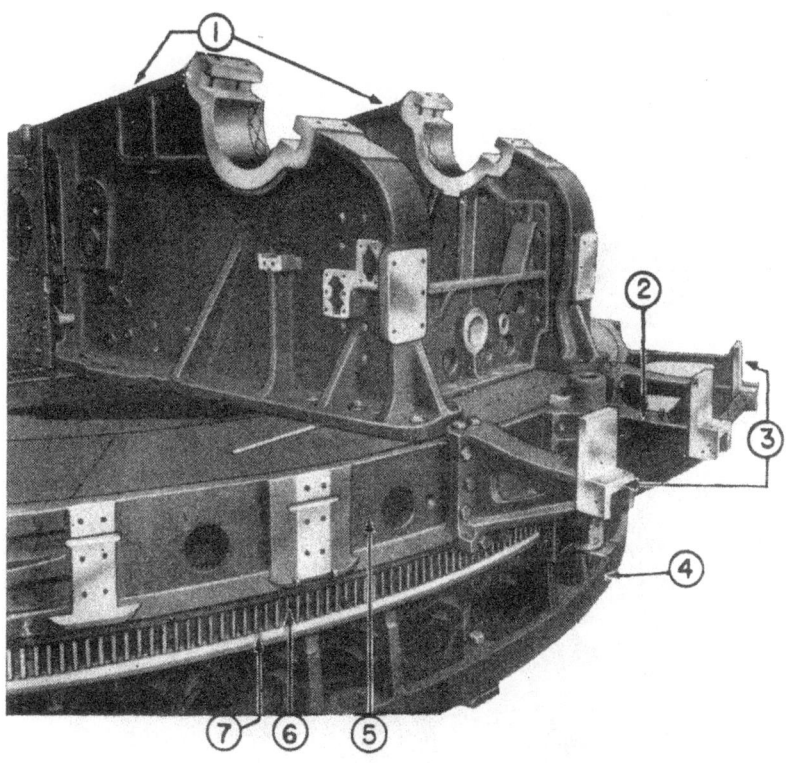

1. Side frames.
2. Traversing bracket.
3. Platform brackets.
4. Base ring.
5. Racer.
6. Traversing rack.
7. Azimuth circle.

Figure 82. 16-inch gun barbette carriage M4.

(2) *Elevating brakes.* Elevating brakes are employed to retain the gun at any desired elevation and to prevent rotation of the tipping parts during recoil. The brakes must be released before the elevating mechanism is used, since they are normally locked. There are two brakes, one on each side of the carriage, of the drum-and-band (automotive) type. They are operated by two levers, both on the right side of the carriage.

(3) *Elevating buffers.* The elevating buffers (fig. 85) absorb the shock which results from sudden stopping of the gun and tipping parts when they reach an extreme elevation or depression.

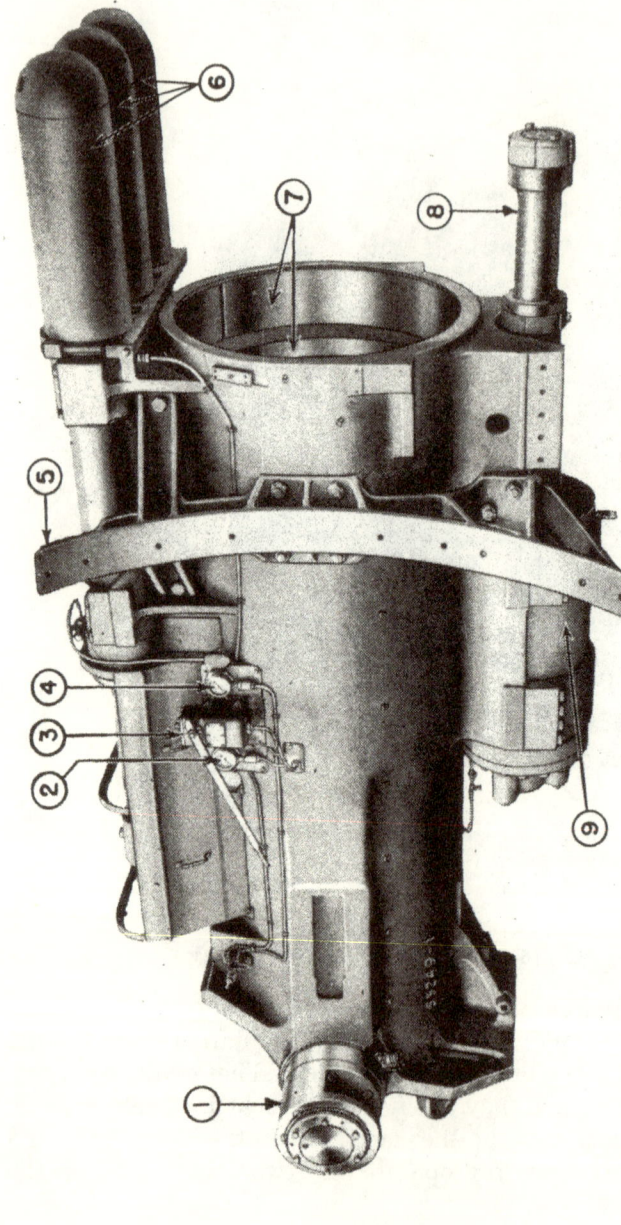

1. Left trunnion extension.
2. Liquid pressure gage and connection assembly, recuperator system.
3. Liquid pump assembly, recuperator system.
4. Air pressure gage and connection assembly, recuperator system.
5. Elevating-rack bracket.
6. Recuperator cylinders.
7. Cradle liners assembly.
8. Recoil piston and piston rod assembly.
9. Recoil cylinder.

Figure 83. Cradle, recuperator plunger top cover and elevating rack removed, 16-inch gun barbette carriage M4.

1. Right elevating gear plate.
2. Right elevating rack.
3. Elevating-clutch lever.
4. Elevating handwheel.
5. Follow-up control handwheel.

Figure 84. Elevating rack, handwheel, and follow-up control, 16-inch gun barbette carriage M4.

These self-contained units are bolted to the frames in such a manner as to make contact with the elevating racks. Buffer levers attached to the buffer housing brackets extend outward in the path of the elevating and depressing stops on the elevating racks. As the rack reaches its maximum limits in either direction, the stops come in contact with the buffer lever and halt the rotation of the tipping parts.

(4) *Antifriction device.* An antifriction device (fig. 86) of the type explained in paragraph 35d (3) is used.

g. TRAVERSING MECHANISM. (1) *General.* The tremendous weight of this large seacoast gun and carriage makes movement in direction a major factor in design. This type of traversing mechanism provides for electric power or hand movement, as the case may require. The traversing bracket, bolted to the racer, houses the pinion and shaft (fig. 87). The pinion meshes with the traversing rack (fig. 80) on the base ring with the result that rotation of the pinion causes the mount to revolve on the conical rollers between the racer and base ring.

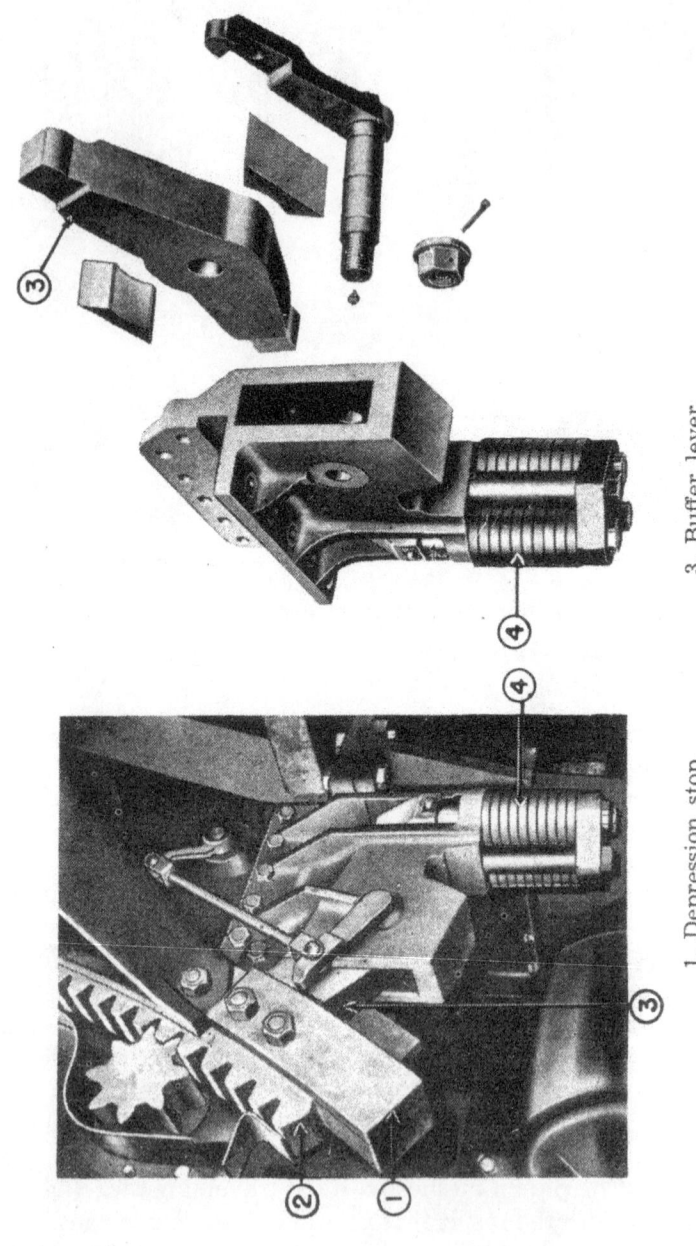

1. Depression stop.
2. Elevating rack.
3. Buffer lever.
4. Buffer Belleville springs.

Figure 85. Right elevating buffer, 16-inch gun barbette carriage M4, assembled and exploded views.

1. Belleville spring rod.
2. Lock nut.
3. Adjusting nut.
4. Belleville springs.
5. Alemite fitting for lubricating the trunnion antifriction roller bearing.
6. Trunnion-roller bearing pin.
7. Main trunnion.
8. Trunnion-elevation pointer.
9. Fulcrum pin.
10. Fulcrum seat.
11. Lever arm.

Figure 86. Right trunnion, elevating scale, and antifriction device, 16-inch gun barbette carriage M4.

(2) *Traversing gear friction device.* A traversing gear friction (overload slip) device relieves excessive strain resulting from sudden starts and stops of the traversing mass and provides positive drive of the traversing pinion within safe limits of strain. The device contains a multi-disk clutch inside the friction box assembly (fig. 87). The grip of the clutch is maintained by the compression of Belleville springs.

(3) *Manual traversing.* Traversing cranks, assembled on crankshafts on the right and left sides of the carriage, are used for rapid change of targets. Accurate adjustment of azimuth is accomplished by using one of two slow-motion traversing handwheels (fig. 87). Clutches, operated by clutch treadles, engage and disengage the traversing slow-motion mechanism.

(4) *Electric power traversing.* To speed up the traversing of the mount and to enable the gun to be pointed in azimuth as soon as

1. Traversing crankshaft assembly.
2. Traversing slow-motion handwheels.
3. Sight mounting brackets.
4. Transverse shaft.
5. Traversing gear-friction box assembly.
6. Traversing pinion.
7. Traversing limit switch.
8. Traversing slow-motion clutch treadle.

Figure 87. Traversing mechanism, 16-inch gun barbette carriage M4.

the target is assigned, a traversing hydraulic speed gear is provided. The traversing pinion may be driven at varying speeds while the motor end of the speed gear is driven at constant speed. Control over power traversing is maintained from one of two control handwheels, one in the azimuth observer's cab and the other at the left-side azimuth operator's station. To cut the power when the mount approaches its limit of traverse, a traversing limit switch is used. This switch breaks the electric current to the traversing motor as the mount approaches the traversing limit in either direction.

h. LOADING MECHANISM. (1) Loading is done by a rammer (fig. 88) operated by either electric or hand power. To load, the gun must be set at the loading elevation, which is approximately $+4°$, and the loading trough extended and lowered to its seat in the breech recess. For electric power operation, a control switch is closed to activate the rammer motor. Because of the great weight of the projectile, the motor must be running at full speed before an attempt is made to ram the projectile. Moving the control lever from neutral to RAM position starts the ramming process which lasts less

Figure 88. Power rammer, overhead trolley, and related parts, 16-inch gun barbette carriage M4.

than 5 seconds. An unstroking device prevents the rammer head from advancing beyond a predetermined point during the ram and withdrawal strokes. The rammer is then withdrawn to repeat the operation in ramming the powder charge into position in the powder chamber. The rammer is controlled by a hydraulic speed gear directly connected to, and driven by, an electric motor. Hand power may be supplied by two cranks, one on each of the right and left sides of the rammer.

(2) Projectiles may be brought to the emplacement from the magazine by means of an overhead trolley and a clamping chain hoist which carry the projectile until it is over the rammer, where it is lowered by the chains onto the rammer trough or parking table. Powder charges are brought up by ammunition trucks. Trucks are also used for hauling projectiles when the emplacement is not equipped with a satisfactory overhead trackage system.

i. AMMUNITION TRUCK M4. The ammunition truck M4 (fig. 89) is provided with aprons extending longitudinally along the sides. These aprons are utilized as bridges to transfer the ammunition from the truck to the rammer trough. Safety dogs on the truck bed prevent the ammunition from rolling during transit and are released by hand for unloading from either side.

Figure 89. Ammunition truck for 16-inch gun barbette carriage M4.

38. Barbette Carriage M1919

a. GENERAL. The chief differences between the barbette carriage M1919 (fig. 70) and the M4 are in the recoil and loading mechanisms. The former mounts a 16-inch Army gun M1919M2 or M3, while the latter mounts a 16-inch gun Mk. II M1 of Navy design.

b. CRADLE. (1) The cradle (fig. 90) is a ribbed casting of considerable complexity. There are two sets of recoil cylinders, each set containing a long- and a short-type cylinder. One set is mounted on the bottom and the other on the top of the cradle. The recuperator system is of the two-cylinder, pneumatic type and is mounted with one cylinder over and one under the gun. A schematic diagram of one set of the recoil and counterrecoil cylinders is shown in figure 91.

(2) *Recoil brake and buffer.* The short cylinders are the conventional hydraulic brakes (par. 25). The long cylinders combine both the recoil and buffer functions. The rear half of each long cylinder is the recoil section (of conventional design); the forward half is the buffer section. In effect, the long cylinders limit the velocity of counterrecoil by the same means the ordinary recoil cylinder employs to limit the velocity of recoil. Small counterrecoil throttling grooves are cut in the walls of the cylinder. As shown in the upper part of figure 91, a sliding valve is mounted on the piston rod. During recoil this valve has no function, but when counterrecoil starts it rides against a buffer piston, closing its orifices and forcing all the liquid to flow through the restricted counterrecoil throttling grooves. This throttling effect limits the velocity of counterrecoil, regardless of the angle of elevation, and insures a smooth and even return to firing position.

(3) *Recuperator cylinders.* Power to return the heavy gun to firing position, regardless of elevation, is furnished by the two pneumatic recuperator cylinders mounted on the cradle. In construction and operation they are the same as the recuperator cylinders on the 16-inch gun Mk. II M1 except that the cylinder contains only one air chamber; that is, there is no check valve (par. 30*b*) to control the velocity of counterrecoil.

c. LOADING MECHANISM. (1) The ammunition supply is handled almost entirely by power. Figure 92 illustrates the general arrangement, and figure 93 shows a loaded projectile car. A circular railroad track is built around the gun emplacement and, by suitable switches, connects with the main track line leading to the magazines. Projectile and powder cars are provided to transport the ammunition from the magazines to the emplacement. The projectile car (9) is secured to the mount by dropping its side rails (8) into recesses in the revolving projectile table (5) and in this position will revolve

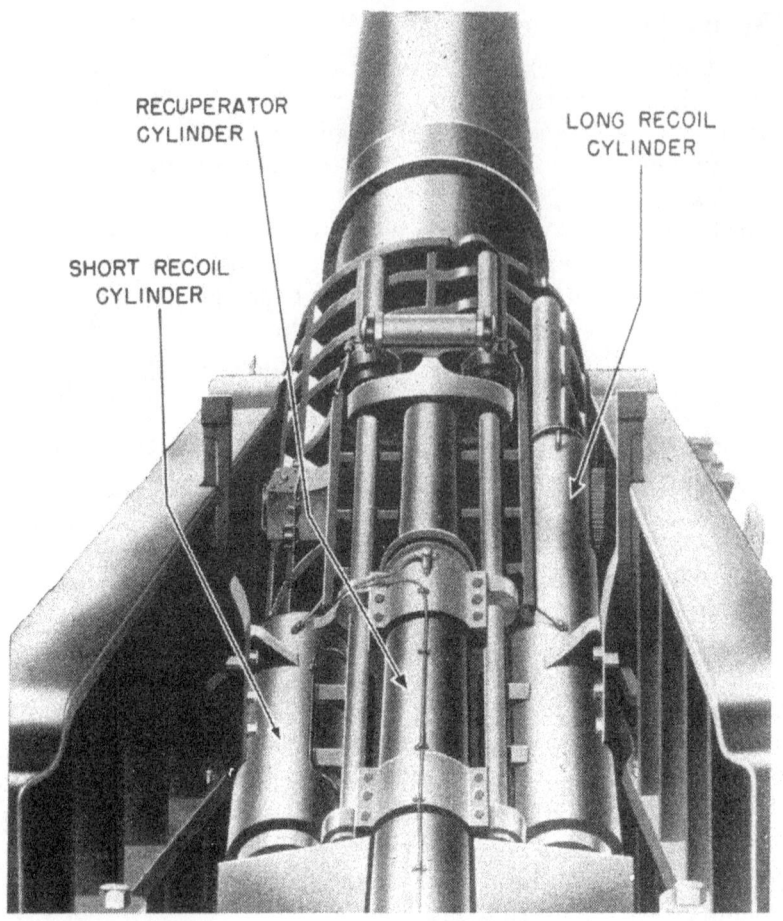

Figure 90. Recoil and recuperator systems, 16-inch gun barbette carriage M1919, top view.

with the mount while being unloaded. The powder car (not shown) coupled to the projectile car may be kept in position to deliver powder, but ordinarily it is taken away as soon as a complete charge has been placed on the powder tray (12). This obviates the danger of having several charges of powder too close to the breech in case of accident. The projectiles are rolled from the car over the side rail bars (8) onto the revolving table (5) which holds three projectiles at a time. The car is then uncoupled from the table, the table revolved, and the projectiles rolled onto the parking table (7). Hand-operated lock stops are provided on both sides of the revolving table to hold the projectiles in place and also to lock the revolving table to the parking table during transfer of projectiles. The parking table is slightly inclined, and the projectiles roll until stopped in position by the hand-operated feed stops indicated by

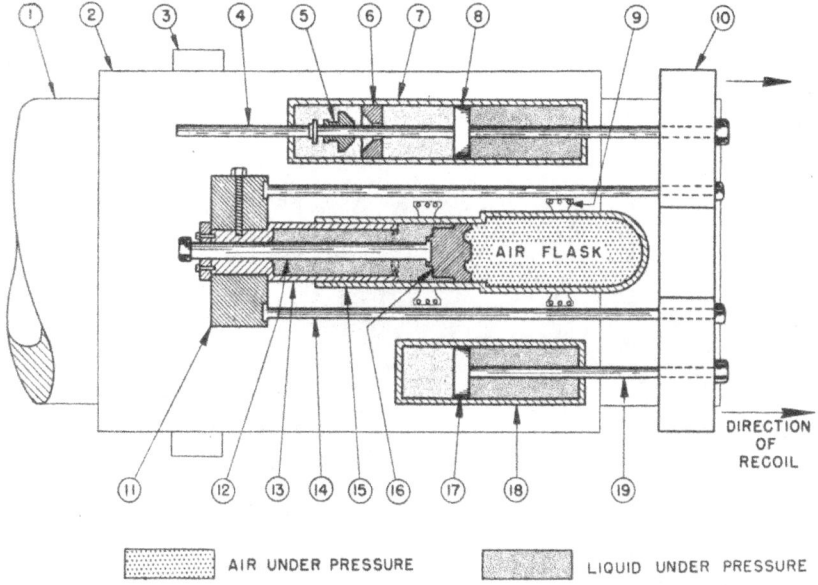

1. Gun tube.
2. Cradle.
3. Cradle trunnion.
4. Recoil piston rod.
5. Sliding valve, counterrecoil buffer.
6. Buffer piston.
7. Long recoil cylinder.
8. Recoil piston.
9. Mounting strap.
10. Recoil band.
11. Yoke.
12. Floating piston rod.
13. Hollow plunger.
14. Connecting rod.
15. Recuperator cylinder.
16. Floating piston.
17. Recoil piston.
18. Short recoil cylinder.
19. Recoil piston rod.

Figure 91. Recoil system, 16-inch gun barbette carriage M1919, schematic diagram.

the small squares. The first projectile lies between these stops. One projectile at a time is fed onto the rammer tray (4), and the spanner tray (2) is lowered into position in the open breech of the gun. The rammer operator at the operator platform (6) controls the rammer (10) by electrical power and pushes the projectile into the gun (1). As soon as the rammer has been withdrawn, the two forward sections of powder (13) are rolled onto the loading tray and shoved into the powder chamber with the rammer. In the same manner the remaining two powder sections are placed in the powder chamber. The spanner tray is now raised and thrown back clear, and the breech is closed ready for firing.

(2) It is possible to have 10 projectiles at the gun at one time; and by coupling 3 powder charge cars to the projectile car, 10 complete rounds can be kept at hand. In considering this loading ar-

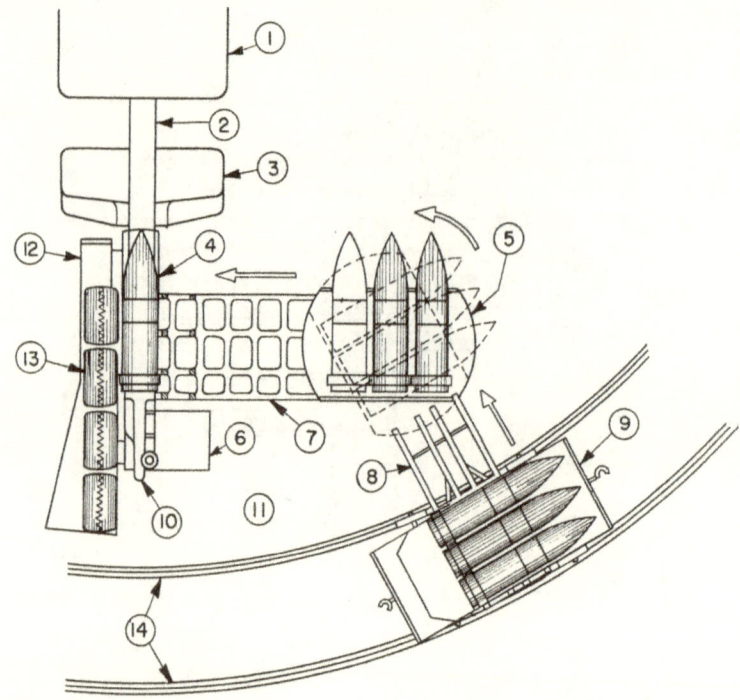

1. Breech of gun.
2. Spanner tray (folds back).
3. Breech-operating platform.
4. Projectile on rammer tray.
5. Revolving projectile table.
6. Operator platform.
7. Parking table, projectiles.
8. Lock bars (side rails of car).
9. Projectile car.
10. Power rammer.
11. Platform on carriage racer.
12. Powder tray.
13. Powder charge on receiving table.
14. Circular track about gun platform.

Figure 92. Loading mechanism, 16-inch gun barbette carriage M1919, open emplacement.

rangement, it is well to remember that the projectile weighs approximately 2,400 pounds and each of the four sections of powder charge weighs approximately 215 pounds. The gun has a loading angle of +4°.

(3) The same loading system is used on the barbette carriage M1920 with the 16-inch howitzer (par. 39). At emplacements where the 16-inch guns M1919M2 and M3 on the M1919 carriage have been casemated, the loading system has been modified accordingly.

39. Barbette Carriage M1920

As a result of the successful use of large caliber howitzers in the European War, 1914-1918, a few 16-inch howitzers (fig. 94) were established in our harbor defenses. However, they are not standard

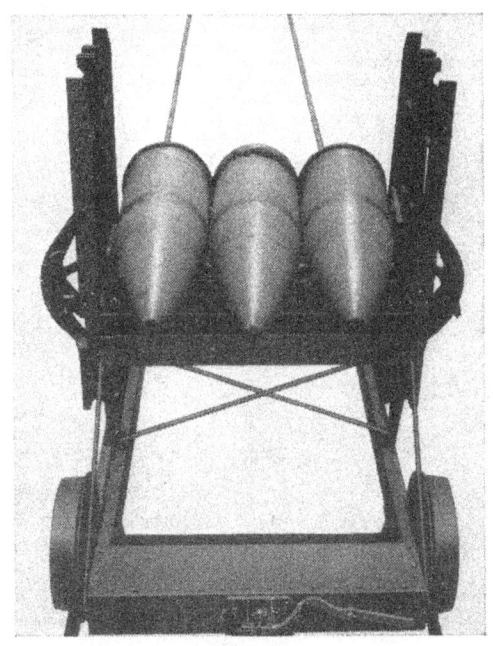

Figure 93. Projectile car, 16-inch gun barbette carriage M1919 and howitzer barbette carriage M1920.

Figure 94. Barbette carriage M1920 with 16-inch howitzer M1920.

armament for future installation. As a result of the decreased weight and power of the howitzer, its carriage is made slightly lighter and simpler than the M1919. It permits all-around fire at elevation between −7° and +65°. The recoil brake consists of four cylinders of the same type as the two short recoil cylinders of the M1919 carriage. However, these cylinders incorporate counter-recoil buffers of the dashpot type instead of the special buffers in the long cylinders of the M1919. The recuperator mechanism is the same as that on the M1919 carriage, but is comprised of only one cylinder mounted at the bottom of the cradle. The loading arrangements and other important features of the carriage are essentially the same as those of the M1919 (par. 38).

40. Barbette Carriage M1917

Although this model is no longer manufactured for our defense installations, it is well to discuss it briefly since many carriages are still employed as long-range mounts for the 12-inch gun M1895M1A4 (fig. 95). The model provides for a 360° traverse and elevations from 0° to 35°. The elevating mechanism, like that of the 16-inch gun carriage, is operated by electric power through a Waterbury hydraulic speed gear. The carriage is traversed by hand. A single conventional recoil cylinder (par. 25), equipped with a dashpot buffer, is mounted on the bottom of the cradle. Four spring-type recuperators (par. 27b) furnish the power to return the gun to battery after firing. The elevating mechanism is of the screw type (par. 35d), and the antifriction elevating device is of the type described in the same paragraph. Most of the carriage is below the level of the floor plate. Figure 95 shows this carriage in an open emplacement. However, such emplacements have subsequently been casemated. A new power rammer (par. 35e) and several additional features have also been provided for this carriage.

41. 8-inch Gun Barbette Carriage M1

a. GENERAL. The 8-inch gun barbette carriage M1 (fig. 97) was designed for use with the 8-inch gun Mk. VI Mod. 3A2 when permanently emplaced. While built with a 360° traverse, the emplacement may limit its traverse to 145° (fig. 96). Its maximum firing elevation is 45°; its minimum firing elevation (usually 0°) depends on the emplacement and the terrain in front of the gun. The carriage permits depression to −5° for loading. There are a few minor manufacturing differences between individual carriages, but they do not affect use and care.

b. BASE RING AND TOP CARRIAGE. This barbette carriage has a base ring and racer (fig. 98) of conventional design upon which a

Figure 95. 12-inch gun M1895M1A4 on barbette carriage M1917.

Figure 96. 8-inch gun Mk. VI Mod. 3A2 on 8-inch gun barbette carriage M1, casemate emplacement.

Figure 97. 8-inch gun Mk. VI Mod. 3A2 on 8-inch gun barbette carriage M1.

Figure 98. Base ring and racer, 8-inch gun barbette carriage M1.

gun-supporting structure is moved in azimuth. A platform (fig. 97) containing the loading mechanism is provided for serving the gun.

c. CRADLE. The cradle (fig. 99) is tubular in shape and is suspended by its trunnions in the side frames of the top carriage. The recoil and counterrecoil systems are housed by the cradle.

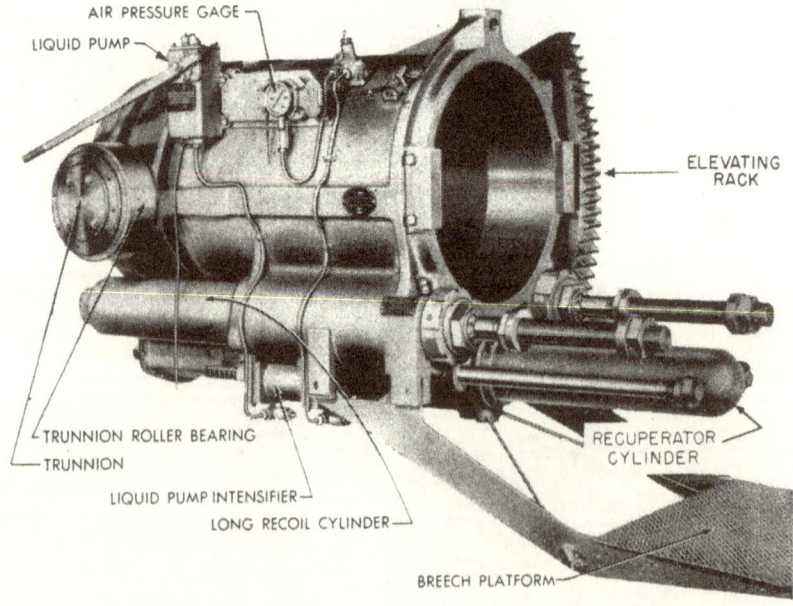

Figure 99. Cradle, 8-inch gun barbette carriage M1.

d. RECOIL SYSTEM. The recoil system (fig. 100) contains two recoil cylinders, one long and one short, attached to the under side of the cradle. The short cylinder is the conventional type with three graduated throttling grooves cut in the walls as described in paragraph 25. The long cylinder contains two pistons on a common piston rod. The rear portion of the long cylinder is practically identical with the short recoil cylinder and its mechanism, but the front portion provides long control over the counterrecoil system. The piston which slides in the front portion of the cylinder has a number of holes. In recoil, these holes offer negligible resistance to the flow of oil. However, during counterrecoil the holes are closed by the action of a spring-loaded valve, and the oil is forced to flow through throttling grooves and thus controls counterrecoil in the same manner as the rear portion of the cylinder controls recoil. The recoil system limits the recoil distance to approximately 27 inches under conditions of maximum powder charge and maximum elevation.

e. RECUPERATOR SYSTEM. The recuperator system (fig. 101) consists of a cylinder and a plunger. The cylinder is attached to the cradle and is filled with air. The plunger is attached to the gun, through the recoil band, by two rods. Packing between the plunger and recuperator cylinder embodies an oil seal. An intensifier (fig. 102) maintains a slightly greater pressure on the oil seal than the pressure on the air in the cylinder. The recuperator unit is normally filled with air at a pressure of 1,600 pounds per square inch. In recoil, this pressure increases to 2,683 pounds per square inch, thereby assisting the recoil system in resisting the backward thrust of the gun and building up energy for the return of the gun to firing position during counterrecoil.

f. INTENSIFIER. The intensifier (fig. 102) performs two functions: it maintains a pressure on the liquid seal of the packing between the recuperator cylinder and plunger, and it facilitates charging the recuperator. The intensifier cylinder is divided into two chambers by a piston. One chamber, filled with oil, is connected to the oil seal in the recuperator plunger packing; the other chamber, filled with air, is connected to the recuperator cylinder. The air pressure in the recuperator, when transmitted to the air chamber of the intensifier, forces the piston against the liquid in the liquid chamber and builds upon the liquid a pressure 14 percent greater than the air pressure in the recuperator. This pressure is transferred to the oil seal. Thus, the greater the air pressure in the recuperator, the greater is the pressure in the liquid seal to prevent the air from escaping.

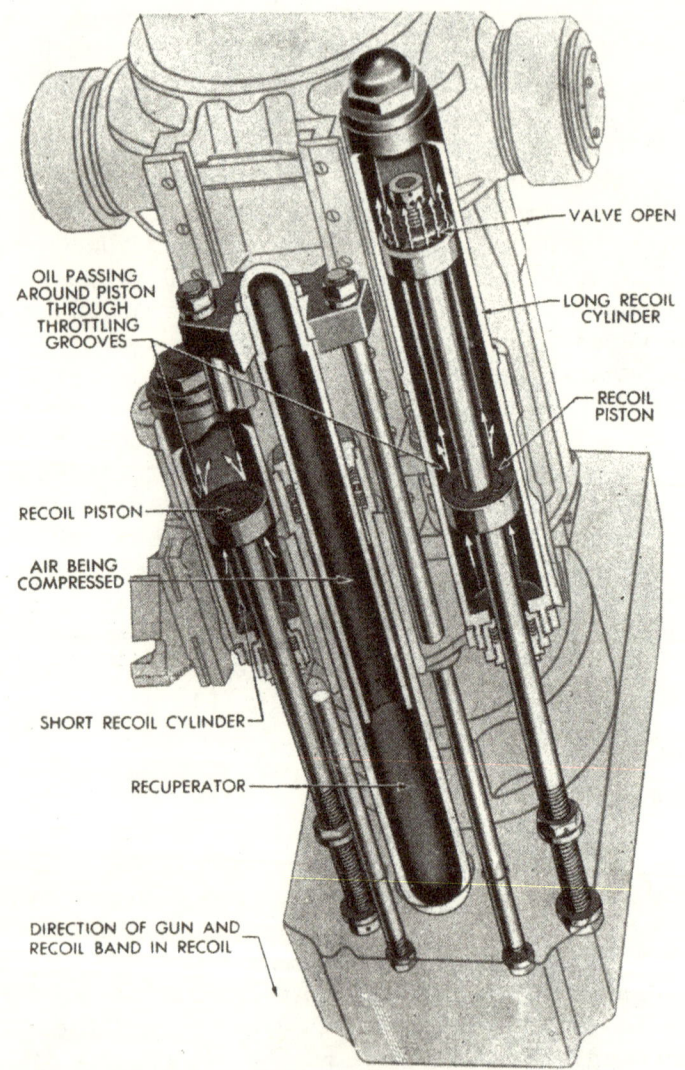

Figure 100. Recoil and counterrecoil systems, 8-inch gun barbette carriage M1, sectional bottom view, gun partially recoiled.

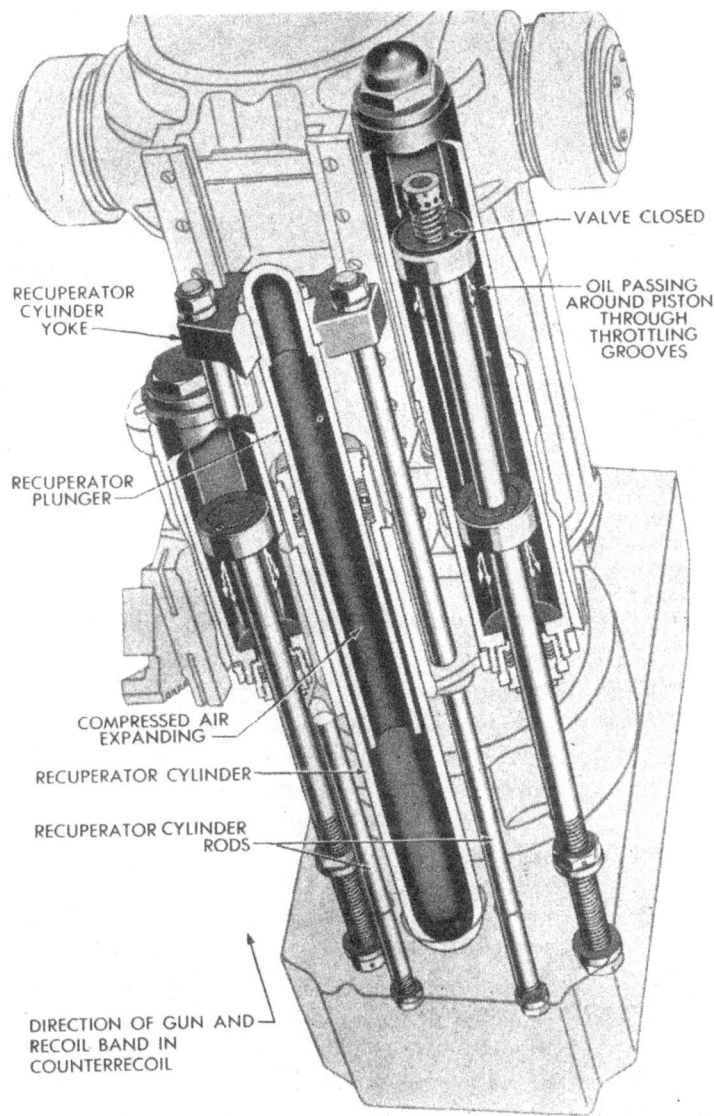

Figure 101. Recoil and counterrecoil systems, 8-inch gun barbette carriage M1, sectional bottom view, gun partially counterrecoiled.

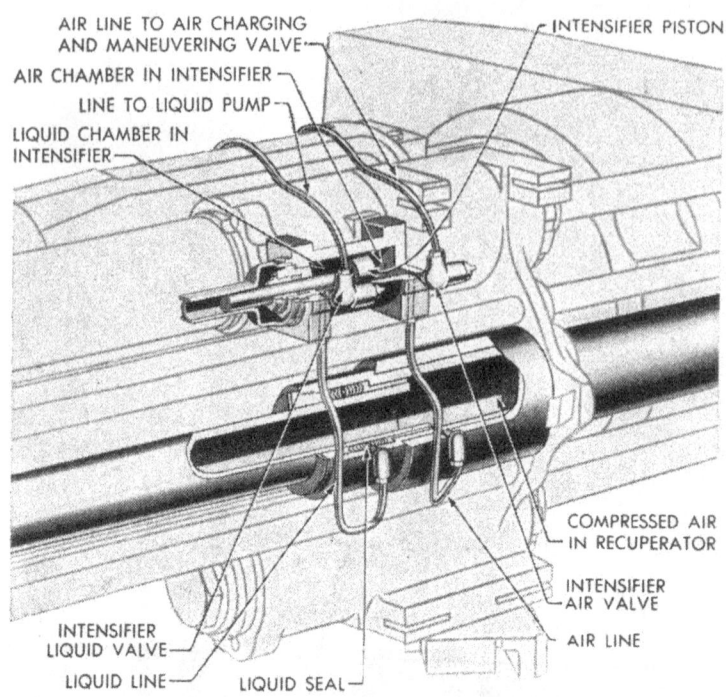

Figure 102. Intensifier, 8-inch gun barbette carriage M1, cut-away view

g. TRAVERSING MECHANISM. The traversing mechanism, located on the left side of the mount, is shown in figure 103. Motion of the handwheels is transmitted by gears and shafts to the worm and pinion mounting in a housing on the inside of the racer. The pinion engages the traversing rack (insert, fig. 103) which is mounted on the inside of the base ring.

h. ELEVATING MECHANISM. The elevating mechanism, located on the right side of the mount, is shown in figure 104. Motion of the elevating handwheel is transmitted by gears and shafts to the elevating pinion which engages and operates the elevating rack on the cradle. Stops are provided at maximum and minimum elevation.

i. LOADING STAND AND TROUGHS. A loading stand (fig. 105) is fixed to the loading platform. It is provided with a folding trough to span the distance between the stand and breech of the gun. Projectiles and powder charges are rammed by hand. The —5° loading depression aids in forcing the projectile home.

j. AMMUNITION TRUCK. Projectiles are brought to the gun from the magazine by overhaul trolley or by an ammunition truck of the type shown in figure 106. A few of these trucks have flanged wheels for operating on rails. The top shelf is for projectiles; the lower shelf is for powder bags. Three projectiles and six powder bags may be carried at one time.

42. 6-inch Gun Barbette Carriage M1

a. GENERAL. (1) The 6-inch gun barbette carriage M1 (fig. 107) is being employed as a mount for 6-inch guns M1903A2 and M1905A2. It is installed in a prepared concrete emplacement which is lowered so that the gun platform is at ground level. A heavy, cast, steel shield, with curved surfaces to aid in deflecting enemy fire, is pro-

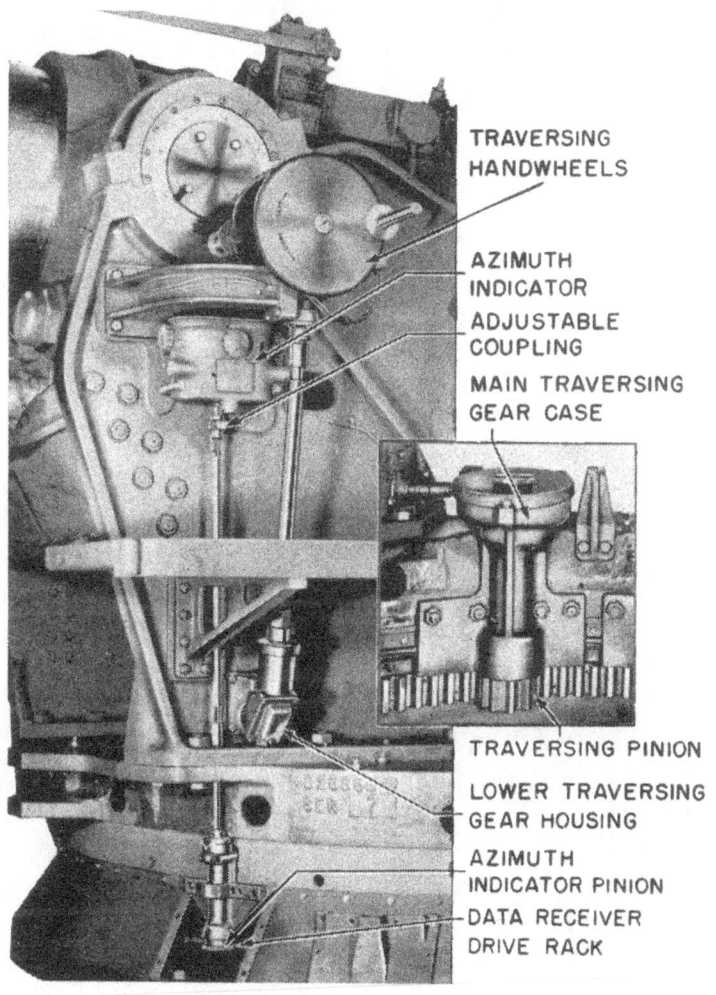

Figure 103. Traversing and azimuth-indicator drive mechanism with insert showing traversing pinion housing on inside of racer, 8-inch gun barbette carriage M1.

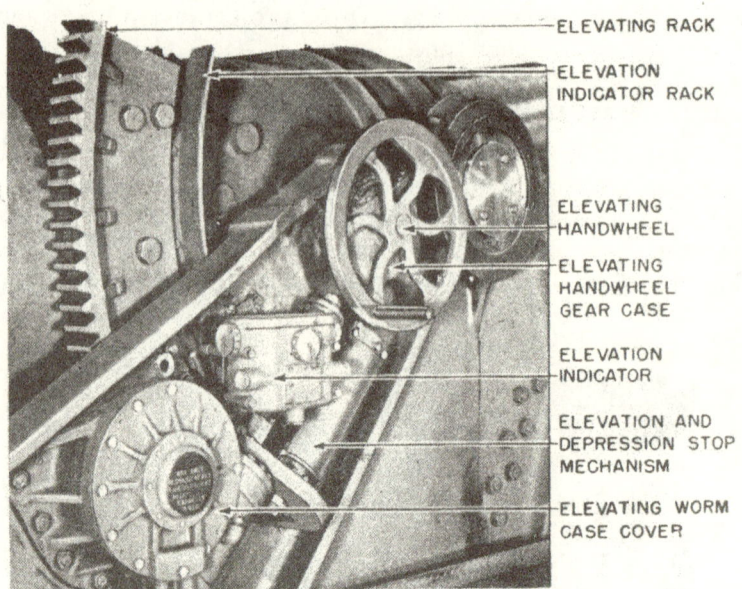

Figure 104. Elevating and elevation-indicator mechanism, 8-inch gun barbette carriage M1.

Figure 105. Ramming the projectile, 8-inch gun barbette carriage M1.

vided to protect the gun crew. As shown in figure 108, the base ring, racer, and top carriage are of conventional design. It has a 360° traverse.

(2) The 6-inch gun barbette carriages M2, M3, and M4 are almost identical with the M1. The chief differences are:

(a) On the M3 and M4 carriages, certain changes in the cradle and in the recoil and counterrecoil systems have been made to compensate for changes made in a 6-inch gun of new design.

(b) The M2 and M4 carriages have a new type of power-driven elevation system (par. 36b). This permits the gun to be automatically set in elevation.

Figure 106. Ammunition truck M17 for 8-inch gun barbette carriage M1.

b. ELEVATING MECHANISM. (1) The gun may be elevated by hand or by electrohydraulic power. The hydraulic system (fig. 109) is powered by an electric motor which drives a constant-speed, variable-delivery fluid pump (A-end) operating a constant-displacement fluid motor (B-end) which is geared to the elevating mechanism (par. 36). When the handwheels are turned, the hydraulic pump (A-end) delivers oil under pressure to the hydraulic motor (B-end) to elevate or depress the gun. By placing the clutch-operating mechanism in the HAND position, the elevating mechanism may be operated by hand power.

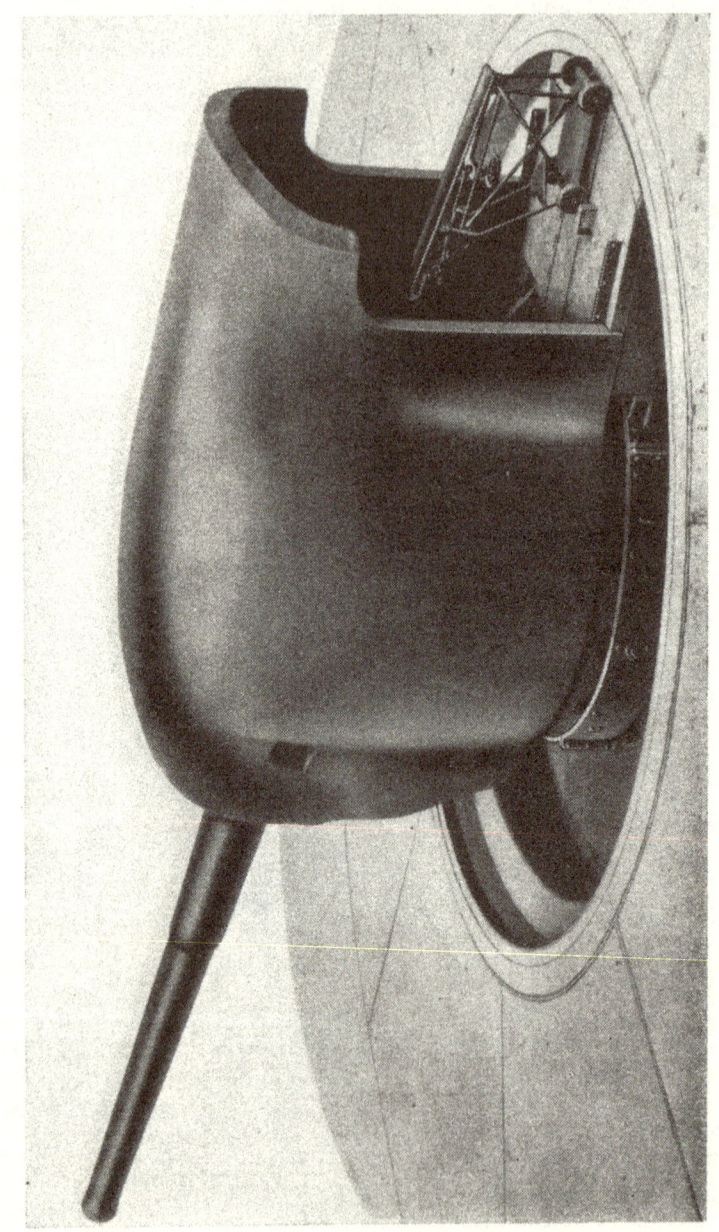

Figure 107. 6-inch gun on 6-inch gun barbette carriage M1.

(2) The power is transmitted (fig. 110) to the elevating rack, attached to the top carriage, as follows:

(a) *Hand operation.* The power is transmitted from the handwheel shaft to the worm drive bevel gears, from the worm drive bevel gears to the elevating worm, from the elevating worm to the worm wheel, from the worm wheel to the elevating pinion gear, and from the pinion gear to the elevating rack (not shown).

(b) *Electric power.* The power is transmitted from the hydraulic motor (B-end) to the power shaft, from the power shaft to the bevel gears, from the bevel gears to the elevating worm, from the elevating worm to the worm wheel, from the worm wheel to the pinion gear, and from the pinion gear to the elevating rack. In electric power operation the handwheels are used to control the variable-delivery fluid pump (A-end) from the handwheel shaft to the compound gear and from the compound gear to the response drive (to signal shaft), which operates the response mechanism of the variable-delivery fluid pump (A-end).

c. TRAVERSING MECHANISM. The mount is traversed by hand power by a direct-drive traversing mechanism (fig. 111). Turning the traversing handwheels rotates the worm, worm wheel, and pinion. The latter engages the traversing rack, mounted in the base ring, and moves the top carriage to the right or left. The azimuth

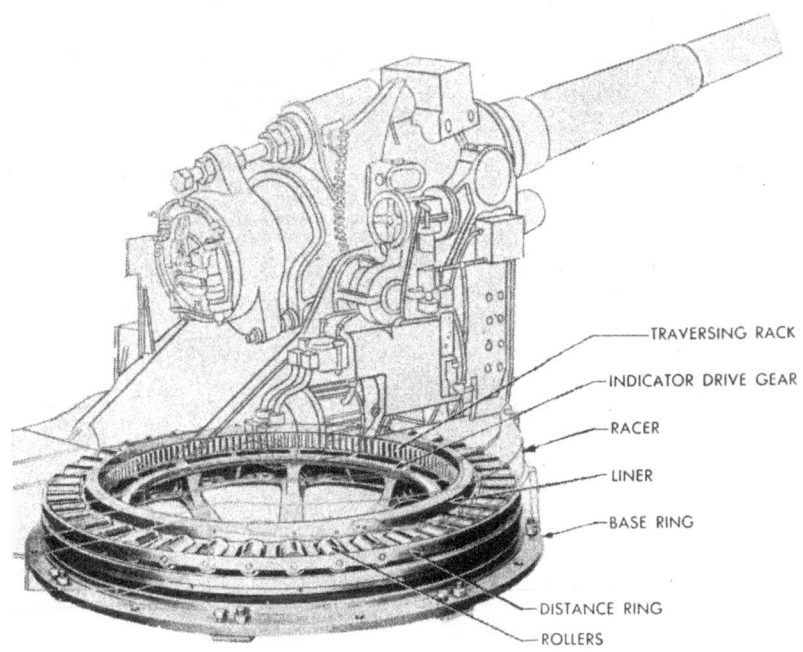

Figure 108. Base ring and rollers, 6-inch gun barbette carriage M1.

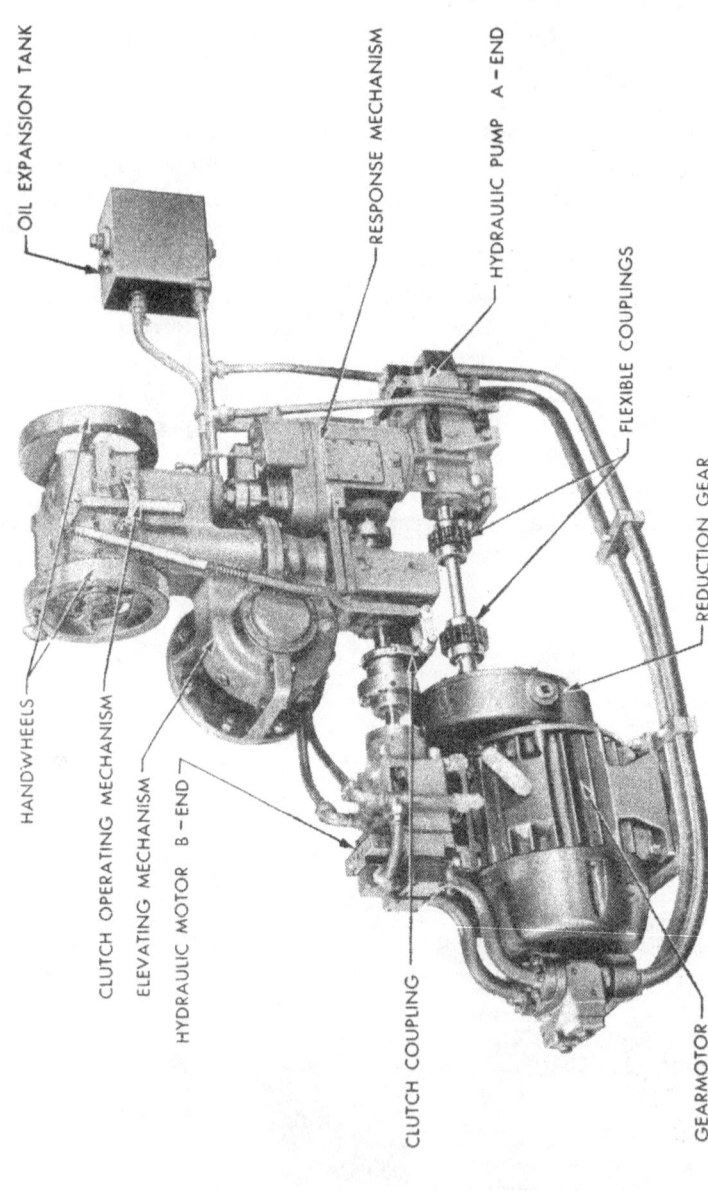

Figure 109. *Elevating mechanism and hydraulic system, 6-inch gun barbette carriage M1.*

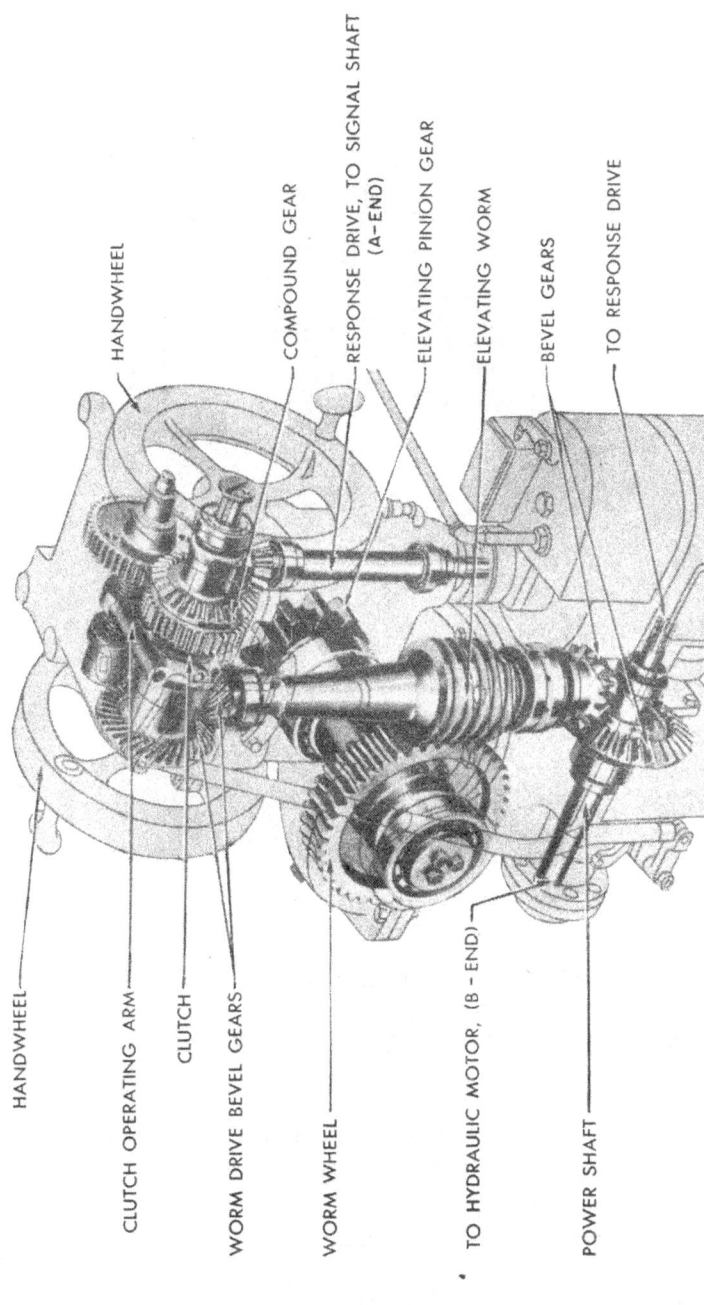

Figure 110. Elevating mechanism, 6-inch gun barbette carriage M1.

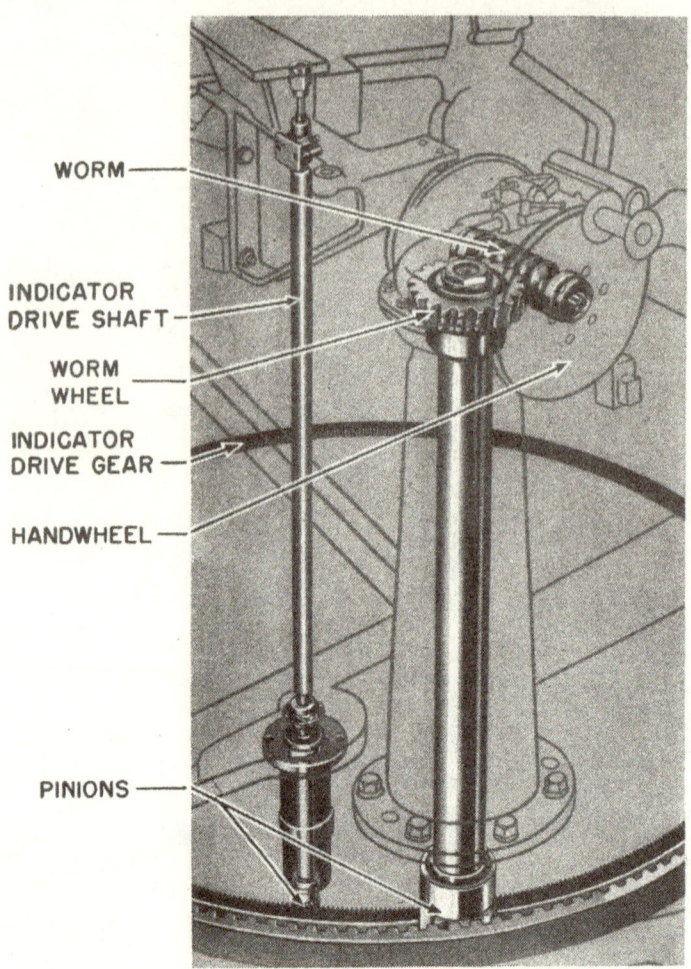

Figure 111. Traversing mechanism and indicator drive, 6-inch gun barbette carriage M1.

indicator is operated by a drive shaft which in turn is driven by a ring gear attached to the base ring.

d. LOADING CARRIAGE. The loading carriage (fig. 112), from which projectiles and powder charges are rammed into the breech, is a four-wheeled truck which is rolled on rails built into the loading platform. A spring and hydraulic buffer, attached to the carriage, bring the latter to a gradual stop when it is rolled against the breech. After loading, the carriage is rolled back and locked by a foot-operated, pedal-released carriage latch. The loading carriage is designed to load a gun elevated to 177.8 mils (10°). However, the loading tray may be adjusted to any elevation between 160.0 mils and 195.6 mils.

Figure 112. Loading carriage, 6-inch gun barbette carriage M1.

e. CRADLE. (1) The cradle is shown in figure 113. It is bronze-lined to provide a bearing surface for the gun in recoil and counter-recoil. Slots are cut in the forward cylinder section for the gun recoil slide keys which prevent the gun from rotating. A counterweight is attached to acquire the desired balance to the tipping parts. There is one recoil cylinder mounted on the top, and two recuperator cylinders mounted on the bottom.

(2) The recoil cylinder (fig. 113) is the conventional type (par. 25) with throttling grooves in the cylinder walls. It contains a dashpot-type buffer. The recuperators are of the spring type (fig. 113). Six springs, three of large diameter and three of small diameter, are used in each cylinder. The springs are kept apart by separators. On recoil, the springs are compressed until they exert a total force of approximately 40,000 pounds.

f. GAS-EJECTION PIPE SYSTEM. The gas-ejection pipe system which passes along the left side of the cradle to the gun breech, provides compressed air to clear the bore of burning fragments, inflammable gases, and smoke after each round is fired.

43. 6-inch Gun Barbette Carriage, Pedestal-type

The 6-inch gun M1900 is mounted on the barbette carriage M1900 of the pedestal type (fig. 76) as described in paragraphs 35b and d. The maximum elevation is about 20°; the minimum elevation, 5°. The recoil system is comprised of one recoil cylinder of conventional

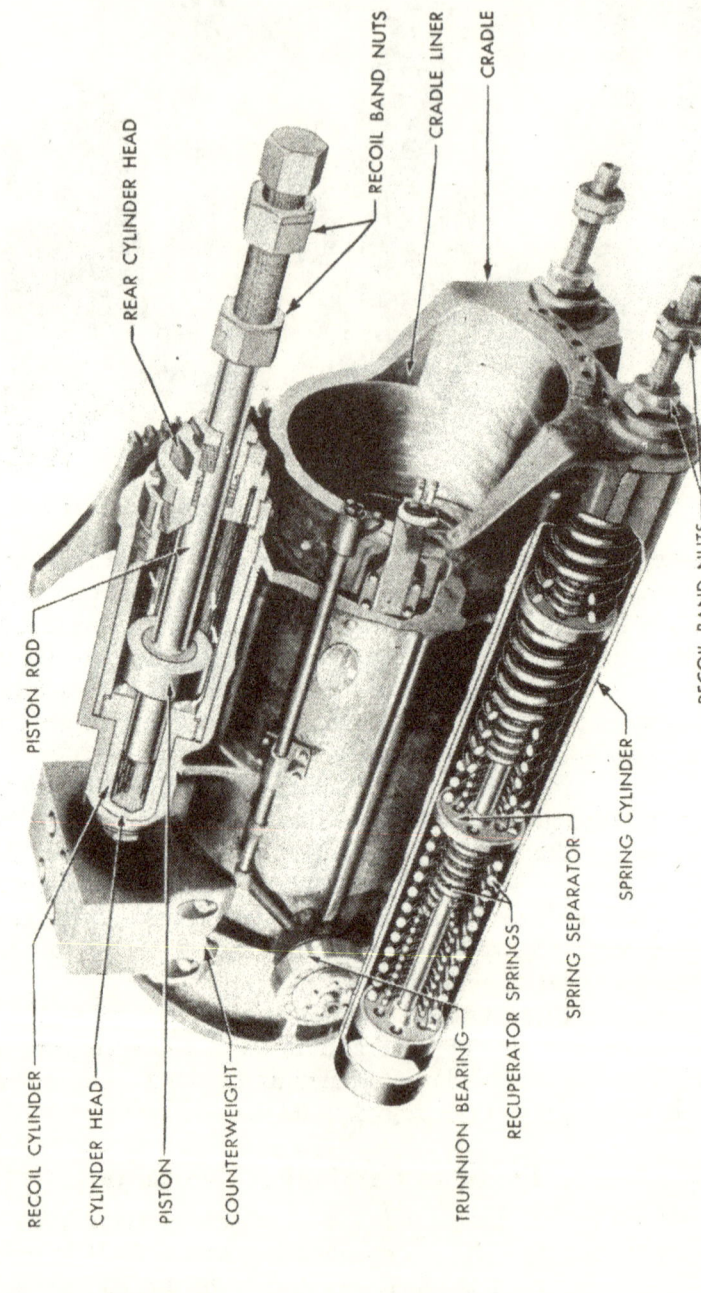

Figure 113. Cradle, with recoil and recuperator cylinders, 6-inch gun barbette carriage M1.

type (par. 25) with throttling grooves in the wall, and two counterrecoil cylinders, each containing four springs with two telescoping into the other two. The buffer system is of the dashpot type located in the recoil cylinder and is similar to that on the 3-inch gun barbette carriage M1903 (fig. 64).

44. 3-inch Gun Barbette Carriages M1902 and M1903

a. GENERAL. Both the M1902M1 and M1903 3-inch guns are mounted on a pedestal mount rigidly bolted to a concrete emplacement (fig. 114). A pivot yoke (fig. 75), which supports the tipping parts, rotates on bearings to facilitate traversing. A shield (fig. 115) is provided to give protection to the gun crew. The general characteristics of this mount are explained in paragraph 35*b*.

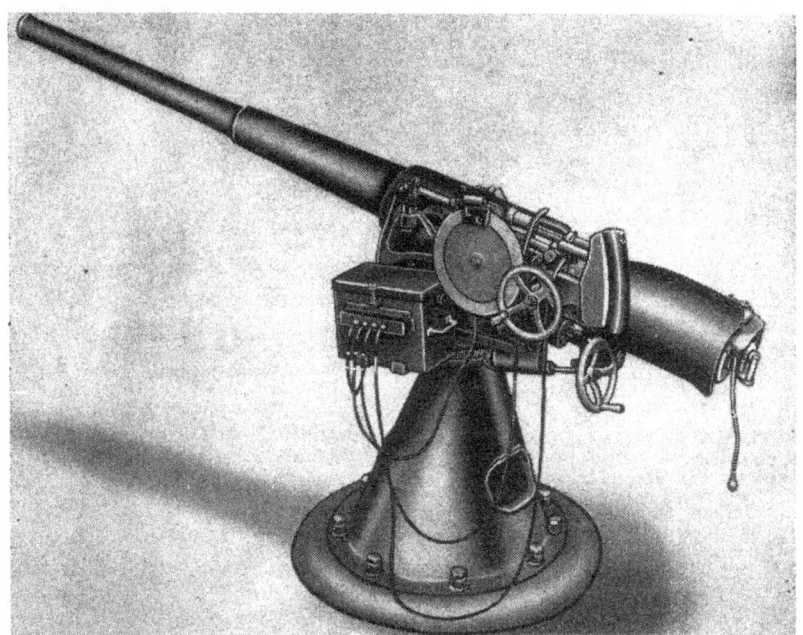

Figure 114. 3-inch gun M1902M1 mounted on 3-inch gun barbette carriage M1902.

b. RECOIL MECHANISM. The hydrospring recoil mechanism of the 3-inch gun barbette carriage M1903 is described in paragraph 29*b*. The recoil system of the M1902 carriage differs from that of the M1903 in that the recoil piston moves to the rear in recoil, instead of the cylinder; moreover, the method of throttling the oil is different (fig. 116). The M1902 system employs a throttling-rod similar to that explained in paragraph 25*c*. Also, the counterrecoil spring is located in the recoil cylinder between the piston rod and cylinder

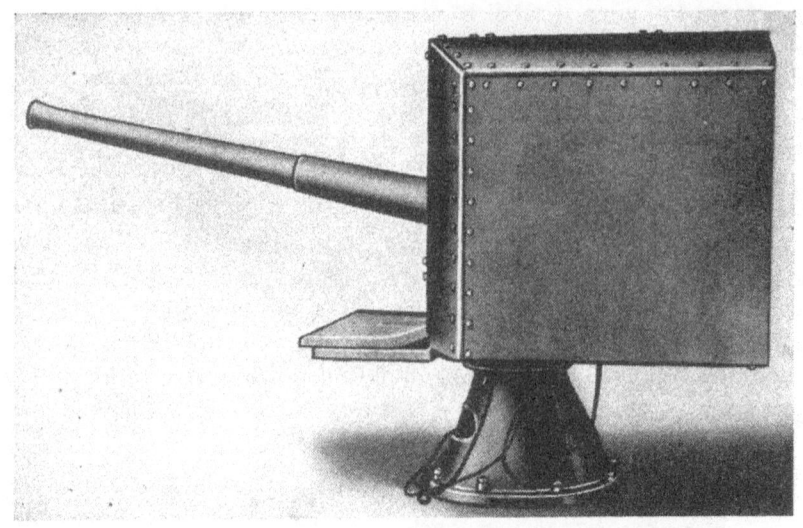

Figure 115. 3-inch gun M1902M1 mounted on 3-inch gun barbette carriage M1902, new-type shield.

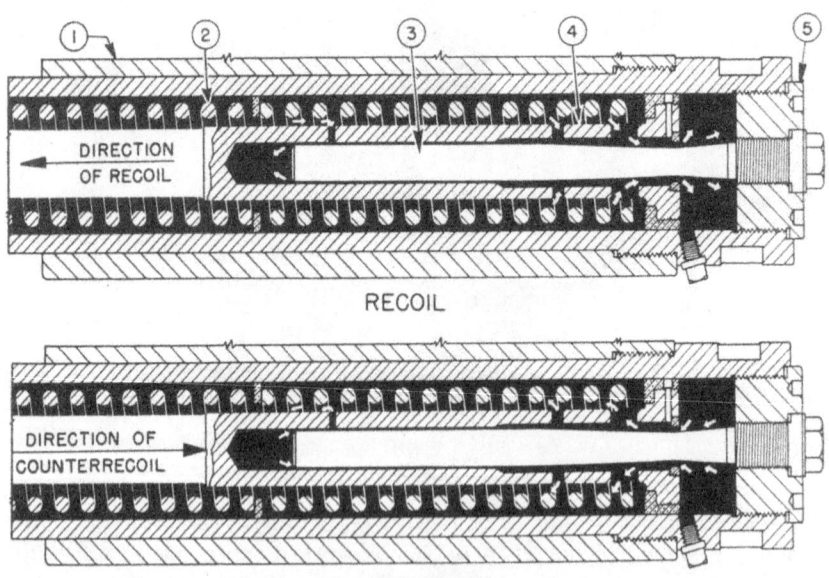

1. Recoil cylinder.
2. Counterrecoil spring.
3. Throttling rod.
4. Piston rod and piston.
5. Cylinder head.

Figure 116. Flow of oil in recoil and counterrecoil mechanism, 3-inch gun barbette carriage M1902.

walls instead of between the recoil cylinder and the spring cylinder as in the M1903 (fig. 64).

c. ELEVATING MECHANISM. The elevating mechanism (fig. 117) of the M1902 carriage is of the rack type. The M1903 elevating mecha-

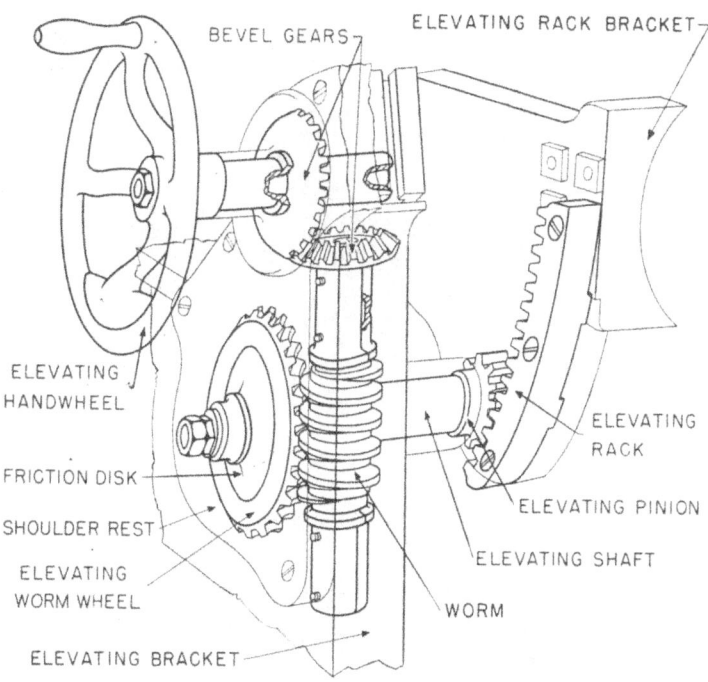

Figure 117. Elevating mechanism, 3-inch gun barbette carriage M1902.

nism (fig. 118) is of the screw type, which operates in the same fashion as an automobile jack as described in paragraph 35d (3).

d. TRAVERSING MECHANISM. The traversing mechanisms of the two carriages are similar. The traversing mechanism of carriage M1903 is shown in figure 119. The traversing rack is held in its position on the pedestal by a friction band. This friction band (overload slip device) permits a certain amount of slippage of the traversing rack, thus preventing injury to the teeth of the rack by the sudden shock of firing. However, the friction band is rigid enough to take the thrust of the traversing worm when the gun is traversed.

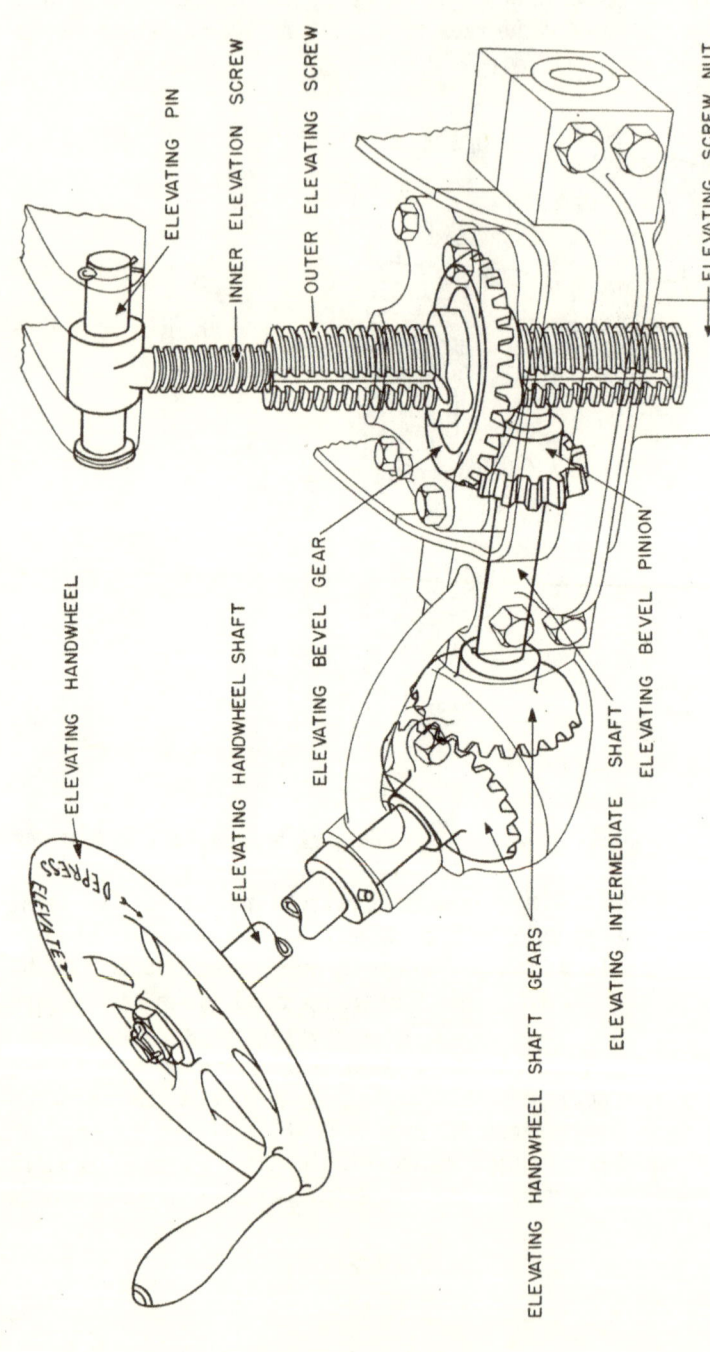

Figure 118. Elevating mechanism, 3-inch gun barbette carriage M1903.

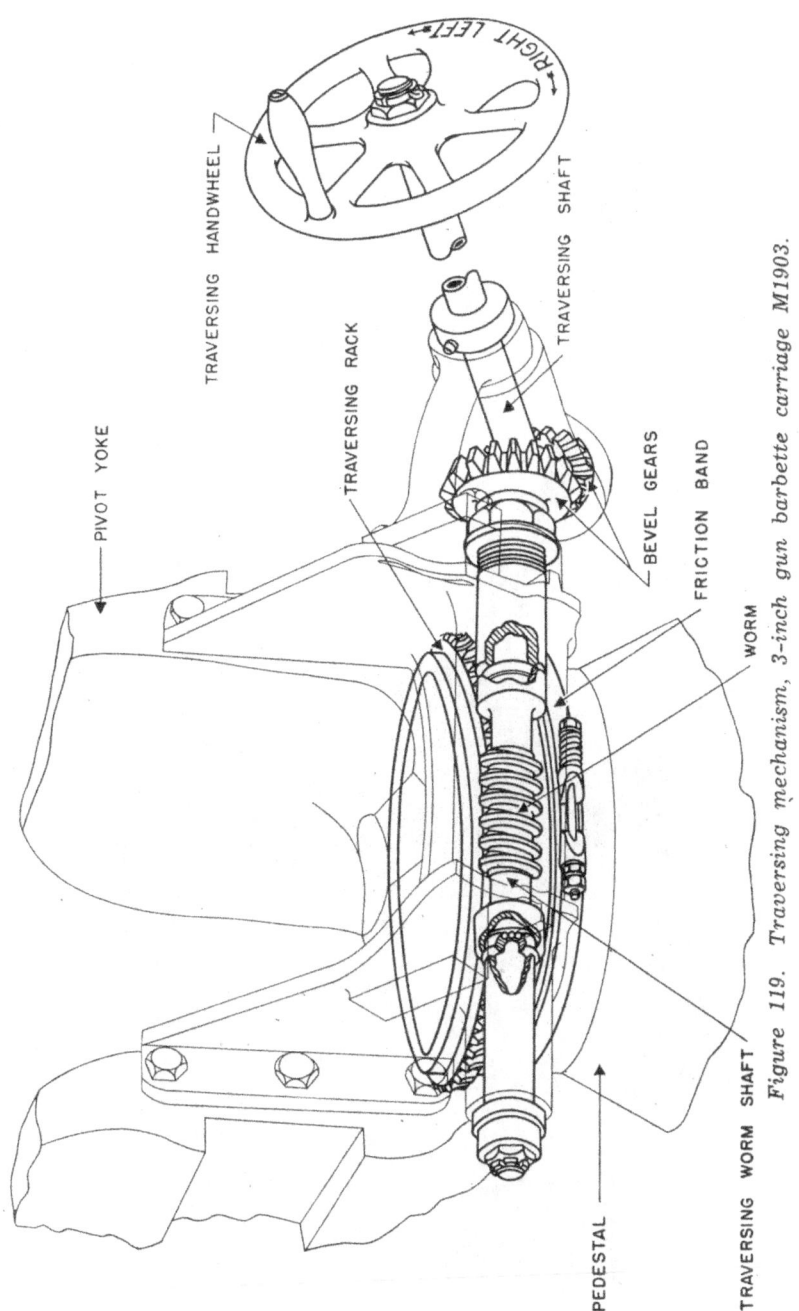

Figure 119. Traversing mechanism, 3-inch gun barbette carriage M1903.

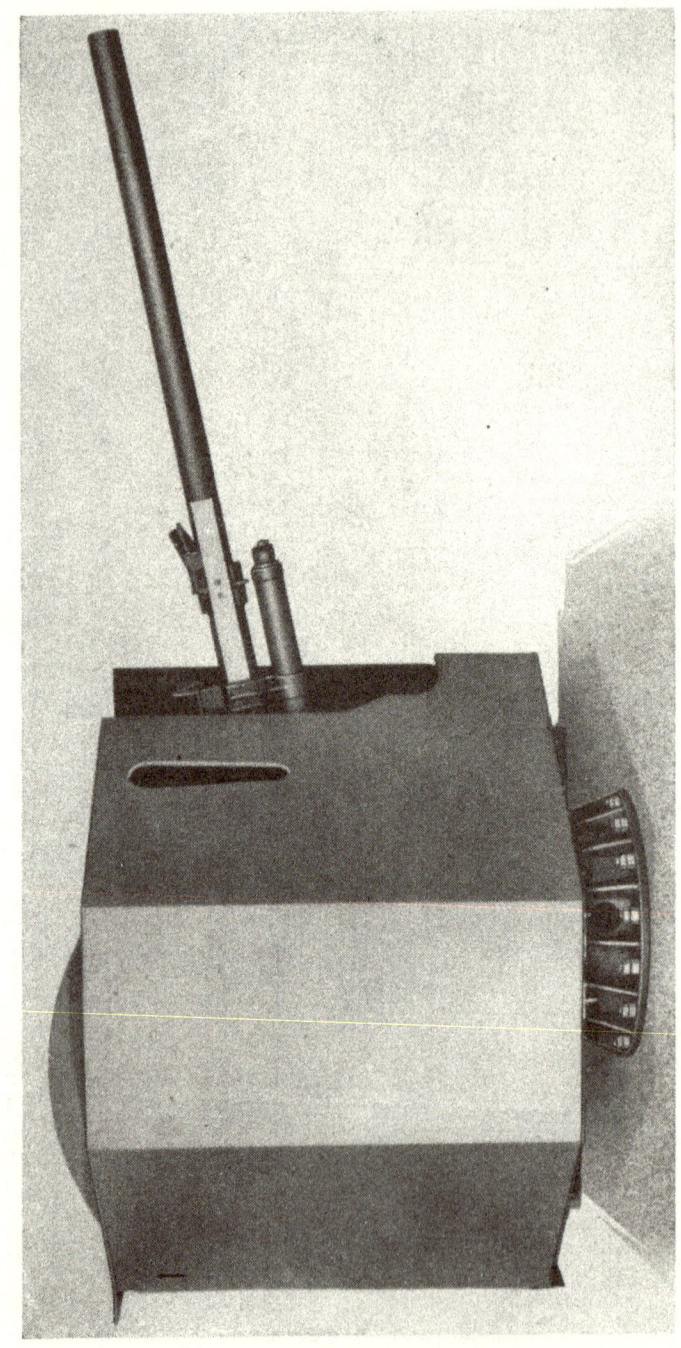

Figure 120. 90-mm gun M1 on 90-mm gun mount M3.

128

45. 90-mm Gun Mount M3

a. GENERAL. (1) Seacoast artillery employs the 90-mm gun M1, supported on the fixed 90-mm gun mount M3, to fire against ground or water targets, especially motor torpedo boats, and for antiaircraft as a secondary mission.

(2) The mount M3 provides for all-around traverse as well as protection for its gun crew and materiel (fig. 120). It consists basically of a pedestal (fig. 121), base ring (fig. 122), top carriage (fig. 123), elevating mechanism (fig. 124), traversing mechanism (fig. 125), equilibrator (fig. 126), cradle and recoil mechanism (fig. 66), and a protective shield and platform assembly (fig. 120). Figure 127 shows the construction of a typical concrete base.

Figure 121. Pedestal, 90-mm gun mount M3.

Figure 122. Base ring, 90-mm gun mount M3.

b. SHIELD. The shield assembly (fig. 120) is attached to the floor plate and has an opening at the rear for loading purposes, and openings in the front for the gun tube and sighting requirements. A diamond-tread, metal flooring supports the shield and provides a working platform for the gun crew.

Figure 123. Top carriage, 90-mm gun mount M3.

c. RECOIL MECHANISM. The recoil and counterrecoil systems of the 90-mm gun mount M3 are described in paragraph 31 (figs. 66, 67, and 68).

d. ELEVATING MECHANISM. The gun may be elevated or depressed by hand or remote control. Two handwheels (fig. 124), working through a train of gears, and an elevating rack provide a means for elevating or depressing the gun by manual operation. An indicator drive assembly registers the elevation or depression of the gun on an indicator dial. A transfer valve permits the operator to change from remote control to manual operation by merely pulling out on the valve and pushing the handwheel and its shaft to the left as far as possible. A reversal of this process returns the mechanism to remote control.

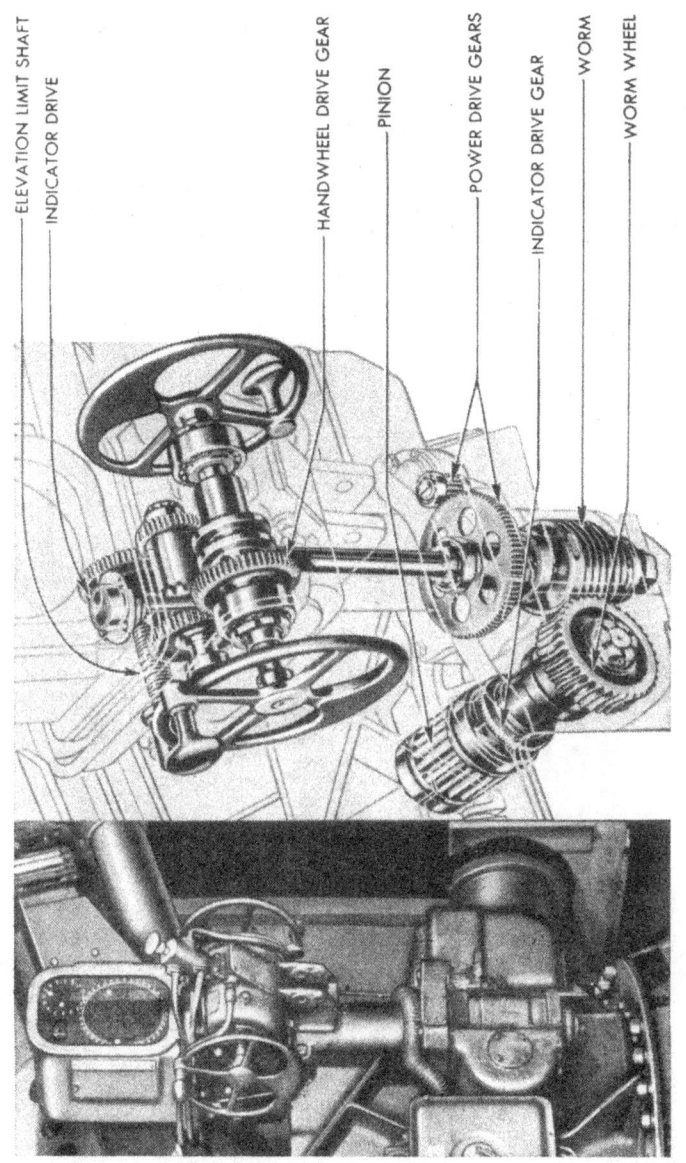

Figure 124. Elevating mechanism, 90-mm gun mount M3.

e. TRAVERSING MECHANISM. The traversing mechanism (fig. 125) is a conventional type equipped with an indicator mechanism which registers the traverse of the gun on the indicator dial. Like the elevating mechanism, it is also equipped with a transfer valve which permits the operator to prepare the mechanism for either remote control or hand operation. By merely pulling the valve out and pushing the handwheel to the left, the mechanism changes from manual operation to remote control. A reversal of this procedure returns the mechanism to hand control.

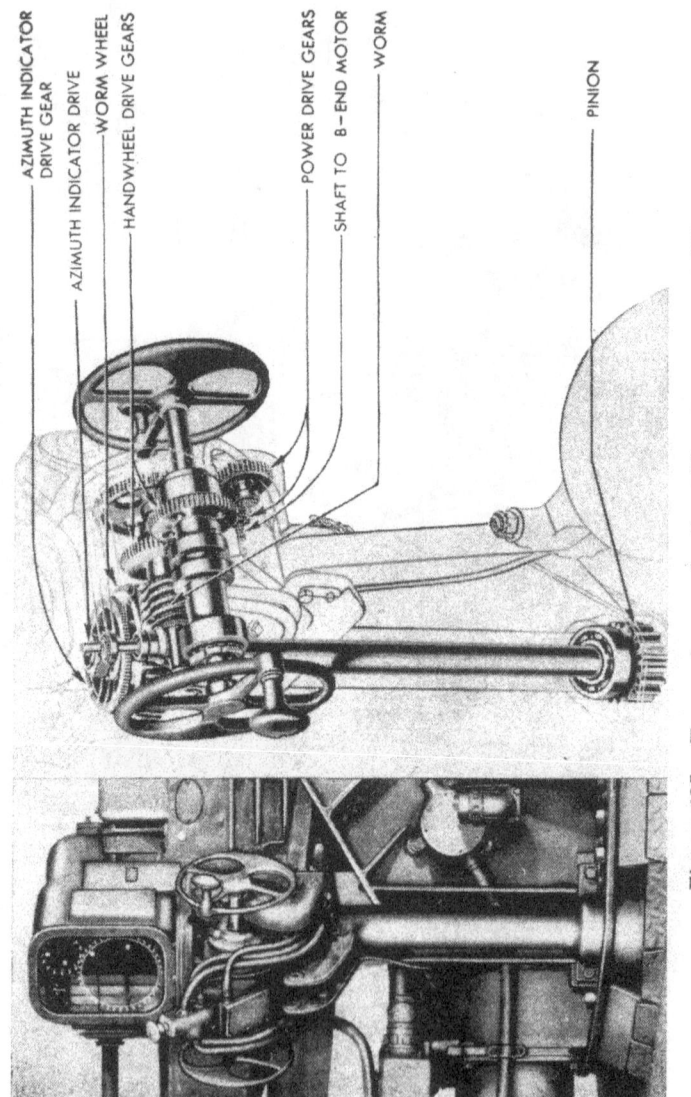

Figure 125. Traversing mechanism, 90-mm gun mount M3.

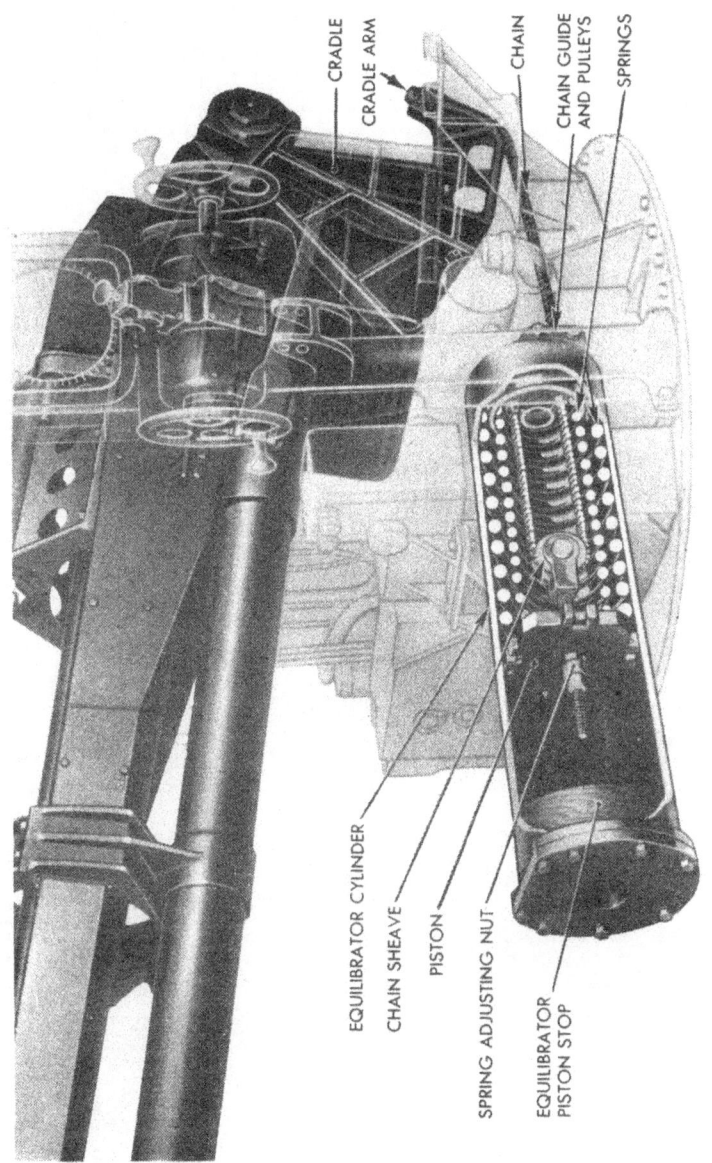

Figure 126. Equilibrator, 90-mm gun mount M3.

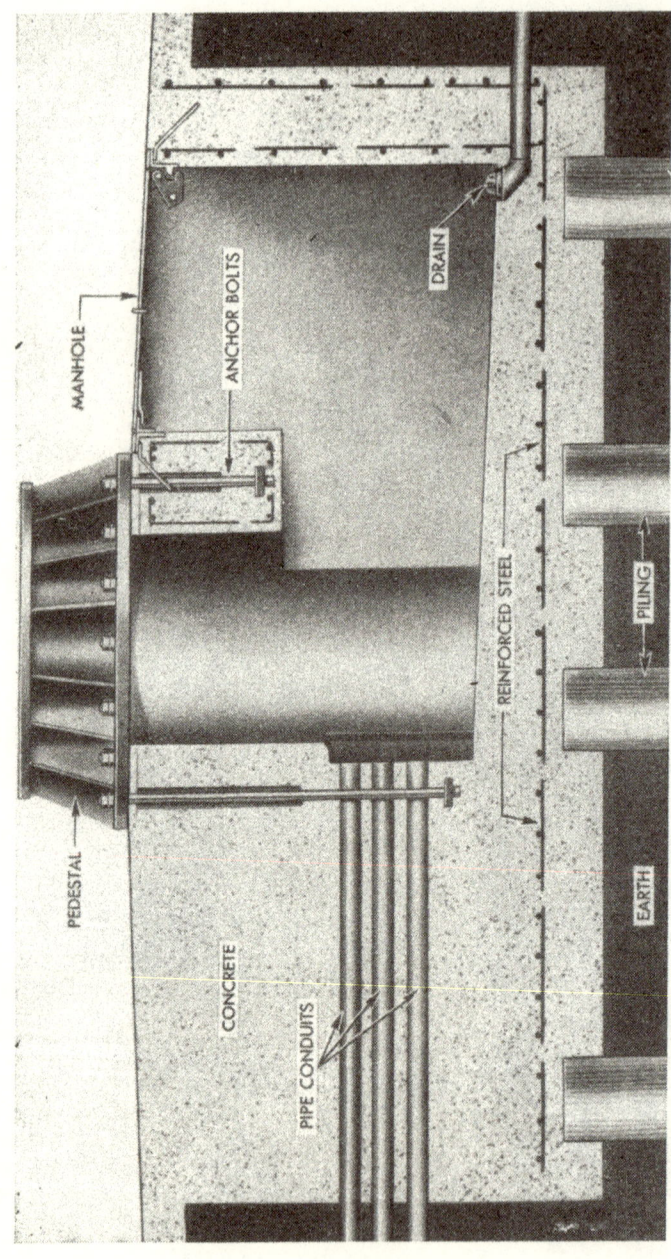

Figure 127. Construction of typical concrete base, 90-mm gun mount M3.

f. EQUILIBRATOR. Because the axis (trunnions), about which the 90-mm gun M1 rotates in elevation, is located to the rear of the center of gravity of the piece, an equilibrator is provided to give balance to the gun and carriage when firing at or near 0° elevation. It also serves to reduce the effort required to elevate or depress the gun. The equilibrator is known as the "puller" type because it pulls downward on the breech end of the cradle rearward of the rotating axis. It is mounted on the left front of the top carriage (fig. 126) and is connected to the cradle by a chain which emerges from the rear of the equilibrator cylinder and passes under the bottom surface of the cradle arm. The front end of the chain passes into the equilibrator cylinder and exerts a pull on the rear end of the equilibrator piston rod. Powerful springs in the rear end of the cylinder force the piston forward to put tension on the chain. At 0° elevation the springs are held at their greatest compression, since at this elevation the cradle arm is at its furthermost position away from the cylinder. As the gun is elevated, the distance between the cradle arm and cylinder decreases, thus releasing the tension on the spring.

Section III. MOTOR-DRAWN CARRIAGES

46. General

During the European War, 1914-1918, the 155-mm gun of French design (M1917 and M1917A1) proved to be a very effective mobile weapon. A new model (M1918M1) of the original was soon manufactured in the United States. Since 1918, further improvements have been made to increase its range and mobility. The M1917, M1917A1, and M1918M1 models with their carriages, because of their French origin, are sometimes referred to as the G.P.F. (Grande Puissance Filloux). Grande Puissance Filloux translated, means a gun and carriage (fig. 128) of great power and the name of the inventor. The present M1 and M1A1 models evolved from this development. Since there are a great number of G.P.F.'s as well as M1's in seacoast defense, each will be discussed in detail.

47. 155-mm Gun Carriage, G.P.F.

a. GENERAL. (1) Carriage models for the G.P.F. are M1917, M1918, M1917A1, M1918A1, M2, and M3. Carriage limber models are M1917, M1918, M1917A1, M1918A1, and the heavy carriage limber M3. They differ very little in general design. All carriages have a single axle, two wheels, a split trail, variable recoil, maximum elevation of

Figure 128. 155-mm gun M1917A1, the 155-mm gun carriage M2, and the heavy carriage limber M3, traveling position.

Figure 129. 155-mm gun M1917A1 and 155-mm gun carriage M2, firing position.

35°, and a traverse range of 60°. All limbers are two-wheeled, designed to support and secure the trails in traveling position, and are provided with a drawbar for coupling to the prime mover.

(2) The M1917 and M1918 models were intended to be moved at very low speeds. They had two steel-bodied wheels, carried on bronze hub liners and equipped with two solid rubber tires. Modified for higher speed, with electric brakes and wheels fitted with antifriction roller bearings, they were designated M1917A1 and M1918A1. Later modifications (steel disk wheels, heavy-duty pneumatic tires, air brakes, and elimination of the original semielliptic spring) led to carriages designated M2 and M3. Inasmuch as the majority of the 155-mm guns have been or will be modified, this manual will cover the 155-mm gun carriages M2 and M3. The modifications of the limber correspond to those made on the carriage. The heavy carriage limber M3 (fig. 130) has steel disk wheels, heavy-duty pneumatic tires, a fifth-wheel type of steering, rigid axle and wheel spindles, and an A-shaped drawbar.

(3) The differences between the M2 and M3 models are minor and do not materially affect troop use and care. These models possess great ruggedness and ease of operation. The firing stresses are transmitted through the trunnions, top carriage, bottom carriage, and trails to the spades which are buried in the ground (fig. 129).

b. CRADLE. The cradle (fig. 131) is pivoted by the trunnions in the trunnion bearings of the top carriage. The gun is carried in the upper part of the cradle where it slides in recoil and counter-recoil. The lower part of the cradle houses the recoil and counter-recoil systems. The elevating sector is bolted to the under side.

c. RECOIL MECHANISM. The recoil system (fig. 132) is of the hydropneumatic, variable recoil type similar to that described in paragraph 26. With this type of recoil system, the length of recoil is automatically shortened as the angle of elevation of the gun is increased. The piston rod of the recoil mechanism is connected to the lower lug of the breech ring and is carried backward with the gun when it is fired. A replenisher cylinder (fig. 133) is connected to the recoil cylinder. It serves as a reservoir for excess oil developed during firing or hot weather. The replenisher contains a piston driven forward by a spring, which keeps the oil in the replenisher cylinder under compression. The piston has an extension prolonged to the rear. This serves as a guide for the piston and as a gage for determining the quantity of recoil oil in the cylinder. The distance from the rear face of the replenisher to the rear end of the piston extension is measured with a rule to determine the amount of reserve recoil oil.

Figure 130. Heavy-duty carriage limber M3, 155-mm gun carriage M2.

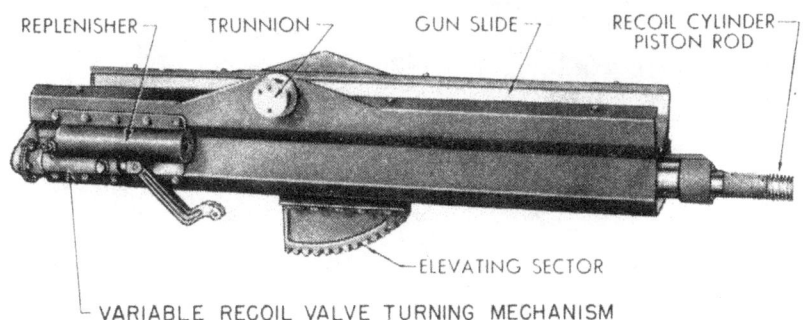

Figure 131. Cradle, 155-mm gun carriage M2.

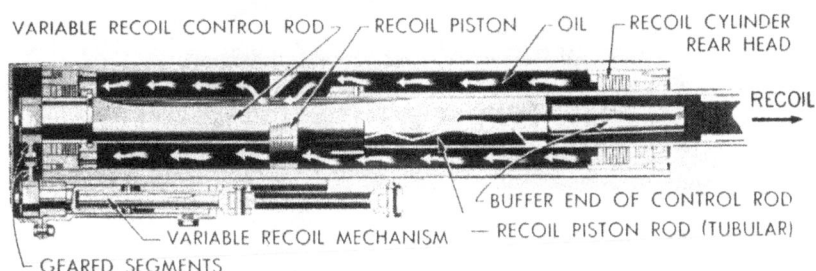

Figure 132. Recoil cylinder and variable recoil mechanism, showing movement of oil when gun is fired, 155-mm gun carriage M2, sectional view.

d. COUNTERRECOIL MECHANISM. The counterrecoil system (fig. 134) is similar to that described in paragraph 29c. The recuperator cylinder is filled with nitrogen at a pressure of 1,592 pounds per square inch at 68° F. The action of the regulator valve is explained

Figure 133. Replenisher and variable recoil valve-turning mechanism, 155-mm gun carriage M2, cut-away view.

in paragraph 28b. It provides long control of counterrecoil. To prevent damage to the regulator valve, a small amount of oil separates it from the floating piston. This oil is known as counterrecoil reserve oil and its presence is indicated by an index in the recuperator cylinder's rear head.

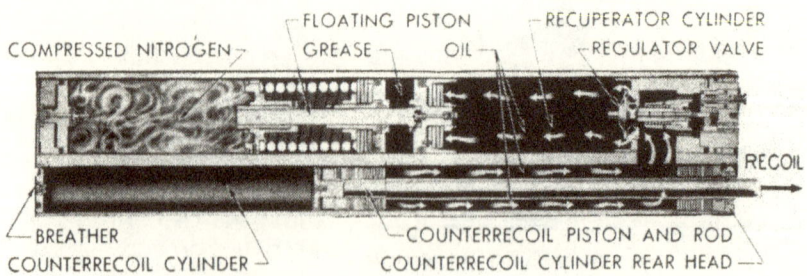

Figure 134. Counterrecoil and recuperator cylinders, showing movement of oil when gun is fired, 155-mm gun carriage M2, sectional view.

e. TOP CARRIAGE. The top carriage (fig. 135) is a large, steel casting mounted on, and secured to, the bottom carriage upon which it traverses. The tipping parts of the mount are supported and pivoted in two trunnion bearings. The handwheels, driving gears, and shafts

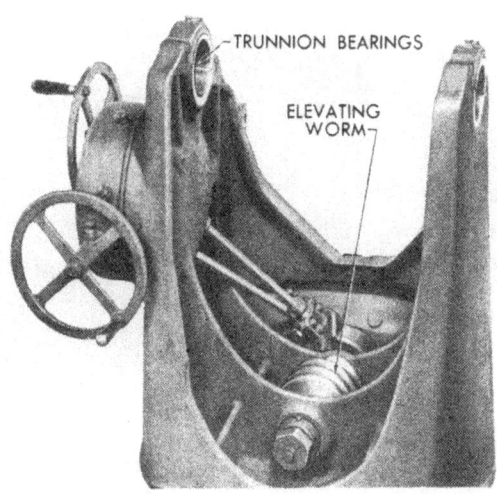

Figure 135. Top carriage showing elevating worm, 155-mm gun carriage M2.

of the elevating and traversing mechanisms are mounted in the top carriage. To facilitate traversing when the gun is at rest, an antifriction bearing (fig. 136) supports the top carriage. This bearing is a small, steel step supported by a column of eight Belleville springs. In firing, the Belleville springs are compressed by the additional downward thrust. The top carriage is then supported by its elliptical

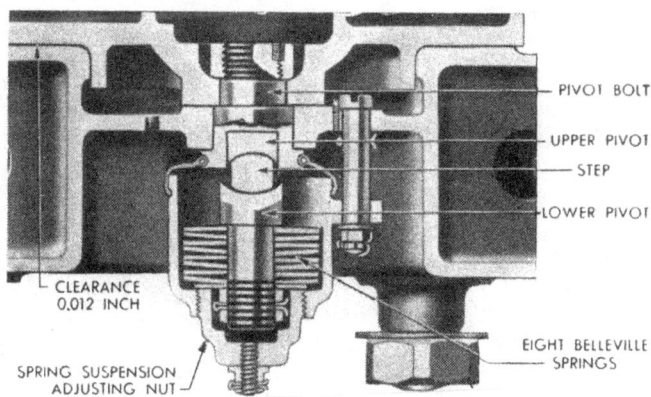

Figure 136. Antifriction mechanism, 155-mm gun carriage M2, sectional view.

141

bearing surface which bears against a similar surface on the top of the bottom carriage (fig. 137). The bringing of the bearing surfaces of the top and bottom carriages into contact, transmits the shock of recoil to the bottom carriage and trails.

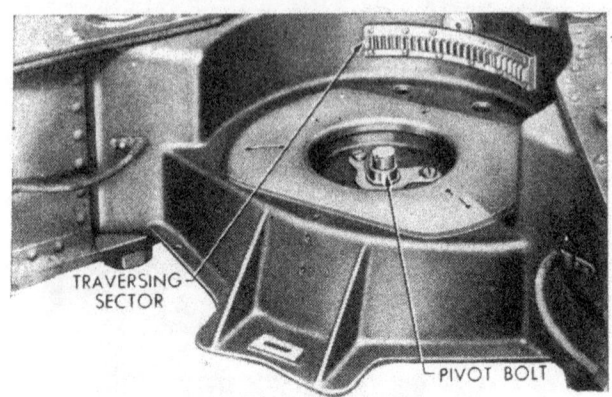

Figure 137. Bottom carriage showing bearing surface for top carriage, 155-mm gun carriage M2.

f. BOTTOM CARRIAGE AND AXLE. The bottom carriage supports the top carriage, provides hinge connections for the trails, houses the antifriction traversing mechanism, and supports an arc-shaped traversing rack. The gun axle bears the weight of the carriage. The axle pivot pin, which holds the axle to the carriage, permits a slight rocking movement of the bottom carriage to compensate for slightly different planes of the wheels when traveling or of the spades when the gun is in battery.

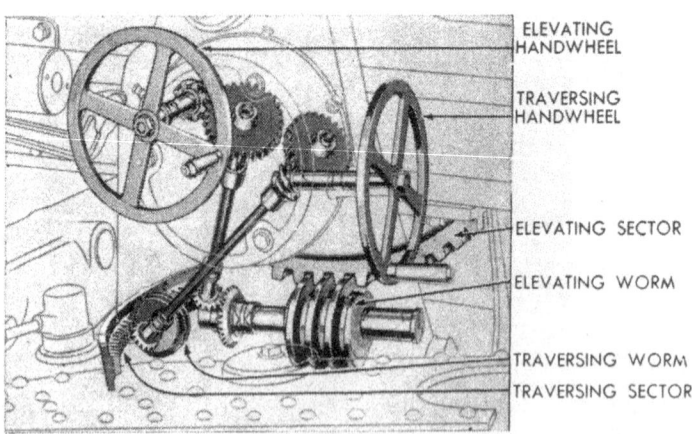

Figure 138. Traversing and elevating mechanisms, 155-mm gun carriage M2, phantom view.

g. TRAVERSING AND ELEVATING MECHANISMS. The traversing and elevating mechanisms each consist of a handwheel and a chain of gears and shafts (fig. 138). The traversing worm engages a traversing sector (rack) on the bottom carriage (fig. 137), and the elevating worm engages an elevating sector (rack) attached to the cradle (fig. 131).

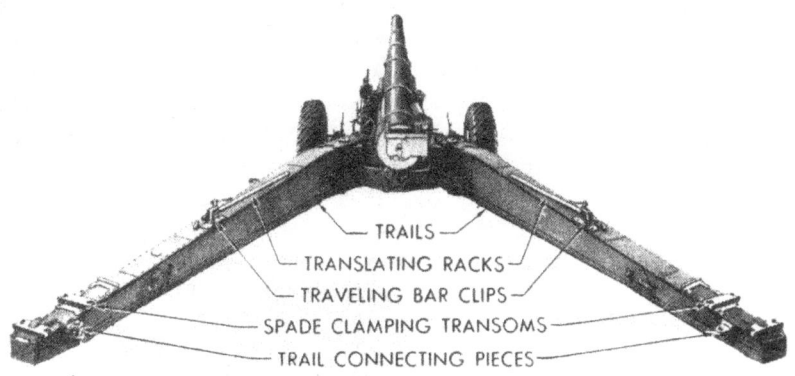

Figure 139. Trails in spread position (without spades), 155-mm gun carriage M2.

h. TRAILS AND SPADES. The two trails (fig. 139), consisting of steel plates and trail ends riveted together to form box beams, are hinged to the bottom carriage by hinge pins. When they are spread, each

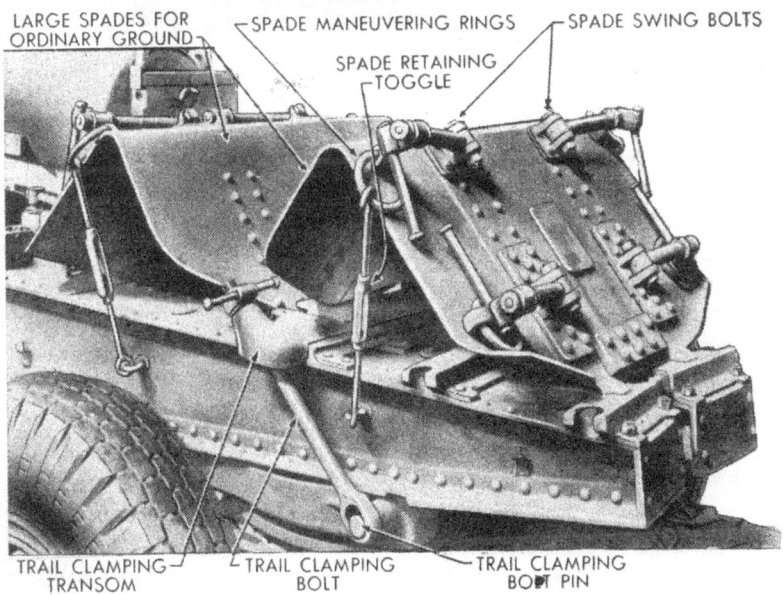

Figure 140. Large spades in traveling position, 155-mm gun carriage M2.

trail forms an angle of 30° with the center of the carriage. Two pairs of spades (fig. 140) are provided, one for soft ground and the other for hard ground.

i. WHEELS AND TIRES. The wheels (fig. 128) are of steel-disk type with 16-ply, heavy-duty tires and bullet-resisting inner tubes. There are two tapered roller bearings in each hub.

j. BRAKES. The brakes on the M2 and M3 carriages are operated by compressed air. The brake system may be considered as consisting of two parts: (1) the air-actuating system, and (2) the brake-operating mechanism. The first is located on the prime mover, the second on the carriage. The brakes are of conventional design and may be operated by hand levers (fig. 141). There is a lever for each carriage brake.

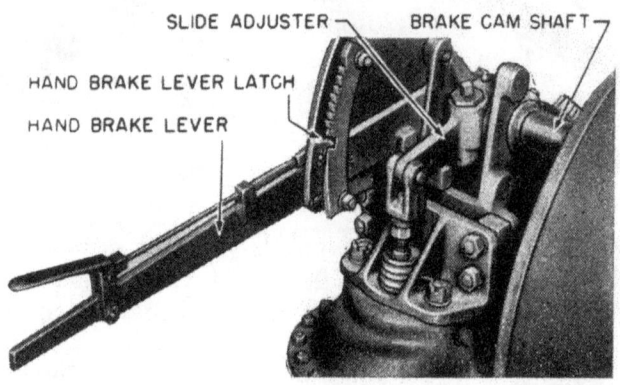

Figure 141. Hand brake lever and connections, 155-mm gun carriage M2.

k. PANAMA MOUNTS. (1) Since the 155-mm gun carriages are designed for a maximum traverse of 60°, a means has been provided to increase the traverse and to enable the gun to act more effectively as a seacoast defense weapon. This is accomplished by using a concrete emplacement (fig. 142) known as a "Panama mount." The advantages that this mount offers are: the gun retains its mobility, the emplacement is stable and firm, and the gun is able to engage a moving naval target throughout its assigned field of fire.

(2) The carriage is mounted on this concrete emplacement with the trail ends riding on a circular rail embedded in the concrete. The arrangement facilitates the making of large changes in azimuth and provides necessary stability in firing by taking the firing stresses at the trail ends. A steel curb band sometimes surrounds the raised inner concrete platform and serves as a guide for the wheels of the carriage, preventing the trail plates from binding on the guide rail in traversing. The gun is traversed in the normal manner, and the trails are moved on the rail by pinch bars. The trails are moved

only when it is necessary to keep the target within the gun's field of fire. Panama mounts may be emplaced to provide a 180° or 360° traverse, depending largely upon the mission assigned and the surrounding terrain.

Figure 142. 155-mm gun on Panama mount.

l. TRAVELING POSITION. When it is desired to go into traveling position (fig. 128), the trails are released from the spades, brought together, and supported on the limber. To equalize the weight between front and rear wheels, the recuperator and recoil piston rods are disconnected from the gun tube, and the latter is drawn to the rear by means of translating racks located on the trails. Figure 143 shows how the gun is moved from firing position to traveling position and vice versa by the translating pinions and racks.

Figure 143. Ratcheting the gun into battery, 155-mm gun carriage M2.

48. 155-mm Gun Carriages M1 and M1A1

a. GENERAL. For all practical purposes the M1 and M1A1 models of the 155-mm gun carriages are identical, the differences being minor and chiefly in the manner of manufacture (fig. 144). The carriage is of the split-trail type designed for a maximum elevation of over 63° (fig. 145). The gun is equipped with longitudinal gun slides and bearing strips upon which the gun slides in recoil and in counterrecoil. The recoil and counterrecoil systems are connected to a recoil lug which is formed by an extension of the under side of the breech ring. The variable recoil system and the counterrecoil system are described in paragraphs 26, 28b, and 29c. The carriage consists mainly of the bogie, bottom carriage, top carriage, cradle, recoil mechanism, equilibrators, and trails (fig. 144).

b. EQUILIBRATORS. (1) To obtain the greatest elevation (fig. 145) of this gun and at the same time not interfere with its recoil, the trunnions are well to the rear of the center of gravity. In other words, it is "muzzle heavy." Therefore, equilibrators are used to neutralize the unbalanced weight and reduce the manual effort needed to elevate the gun.

(2) There are two equilibrators of the pneumatic type. They consist primarily of cylindrical cases and plungers. They are filled with nitrogen gas under pressure and are provided with grease seals to retain the gas. At maximum elevation the equilibrators are in a nearly closed position (fig. 145). When the gun is depressed, the plunger is retracted (fig. 146). This motion draws the plunger ahead and further compresses the nitrogen in the equilibrator. There are temperature adjustment scales for adjusting the tension of the equilibrators at various temperatures.

c. CRADLE. The cradle, which houses the recoil, counterrecoil, and recuperator cylinders, is mounted on the antifriction trunnion bearings of the top carriage. The cradle trunnions form a fulcrum for the movement of the gun in elevation. One end of each of the two pneumatic equilibrators is attached to the cradle near its front end. The other ends of the equilibrators are fastened to the trunnion caps on the top carriage. The cradle also supports the elevating arc as well as the replenisher and the linkage of the variable recoil mechanism.

d. TOP CARRIAGE. (1) The top carriage houses the elevating and traversing mechanisms. The trunnion bearings of the top carriage support the trunnions of the cradle in roller bearings.

(2) The top carriage is seated in, and rotates on, the bottom carriage (fig. 148) which forms the pivot for the movement of the gun in traverse. The antifriction traversing device of the 155-mm gun carriage M1 works on much the same principle as that of the 155-mm gun carriages M2 and M3 described in paragraph 47*e*. However, the top carriage is not supported by a steel step as on the 155-mm gun carriages M2 and M3. Rather, when the gun is at rest the top carriage is supported on roller bearings housed between two races (fig. 150) in the roller path of the bottom carriage (fig. 149). These

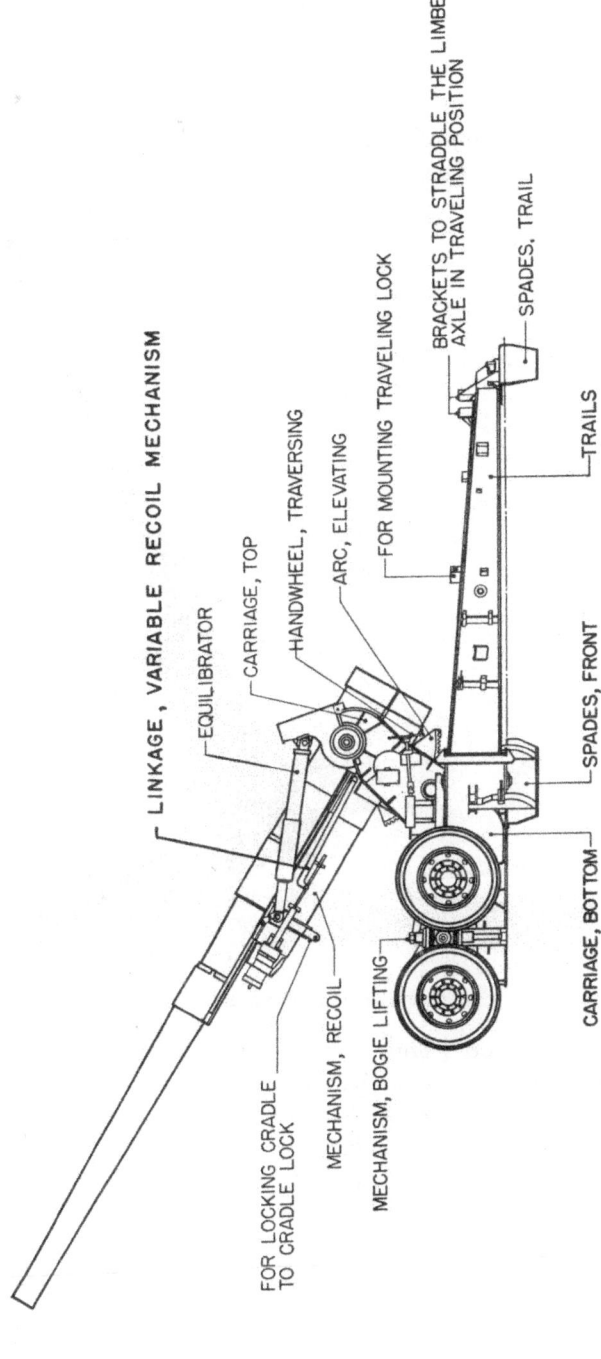

Figure 144. 155-mm gun carriage M1, left side.

roller bearings are in turn supported by Belleville springs (fig. 151). Thus, the full bearing surfaces of the top and bottom carriages are not in contact. When the gun is fired, the Belleville springs are compressed, bringing the full bearing surfaces of the two carriages into contact and enabling the firing stresses to be transmitted to the bottom carriage and trails. The pintle of the top carriage fits into a bushing (fig. 148) in the bottom carriage and is held in place by a special antifriction bolt and nut arrangement.

Figure 145. 155-mm gun carriage M1, firing position, maximum elevation.

e. BOTTOM CARRIAGE. The bottom carriage (fig. 148) forms a support for the top carriage. A traversing rack is bolted to a bracket on the bottom carriage to provide a means for rotating the gun in azimuth. Two trails are connected to the bottom carriage by trail hinge pins. In the firing position, the bottom carriage rests on the ground. In the traveling position, it is suspended on the ends of the bogie-lifting screws.

f. ELEVATING MECHANISM. Motion of the elevating handwheel actuates a series of gears to transmit motion to a shaft and pinion which engages and operates the elevating arc (rack) on the cradle. With

Figure 146. 155-mm gun carriage M1, firing position, 0° elevation.

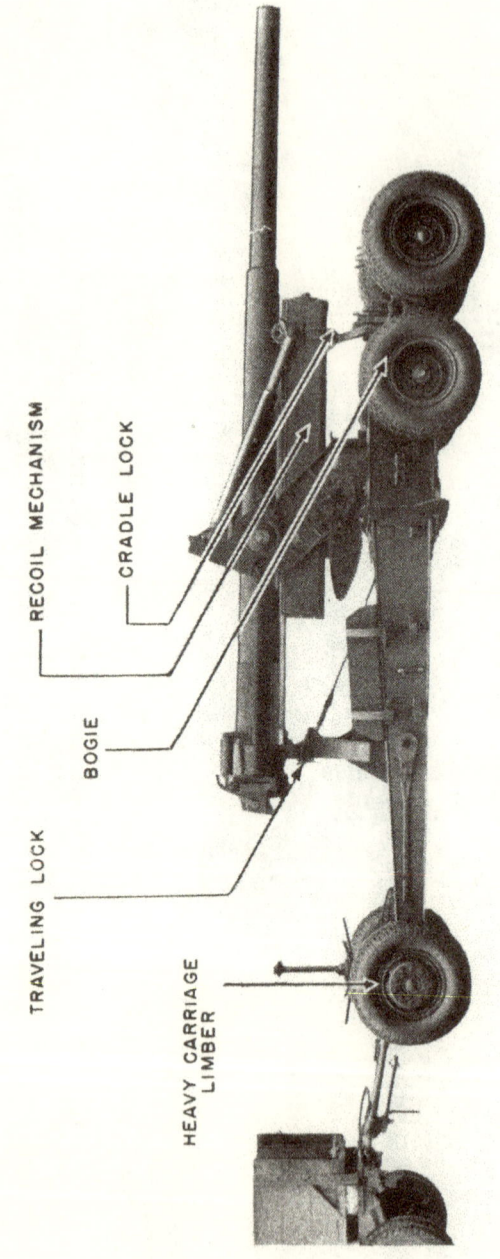

Figure 147. 155-mm gun carriage M1, traveling position.

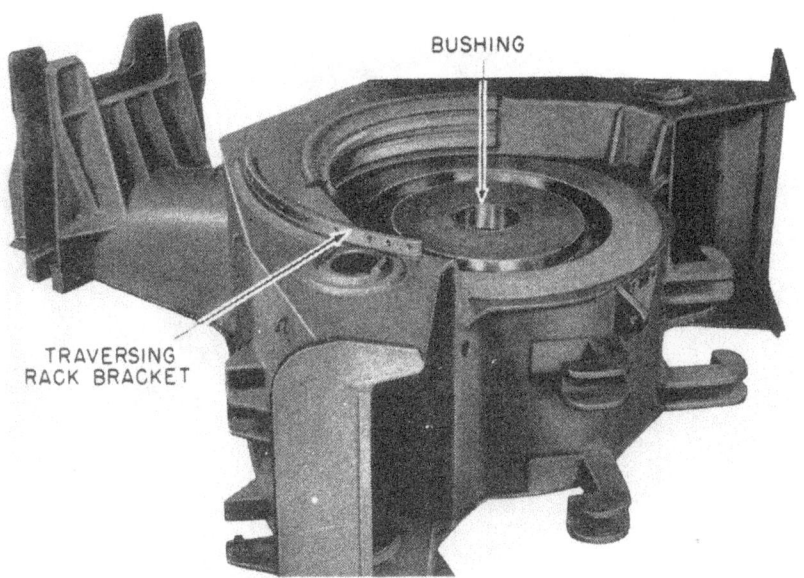

Figure 148. Bottom carriage, 155-mm gun carriage M1.

Figure 149. Roller bearing retainer assembly in bottom carriage, 155-mm gun carriage M1.

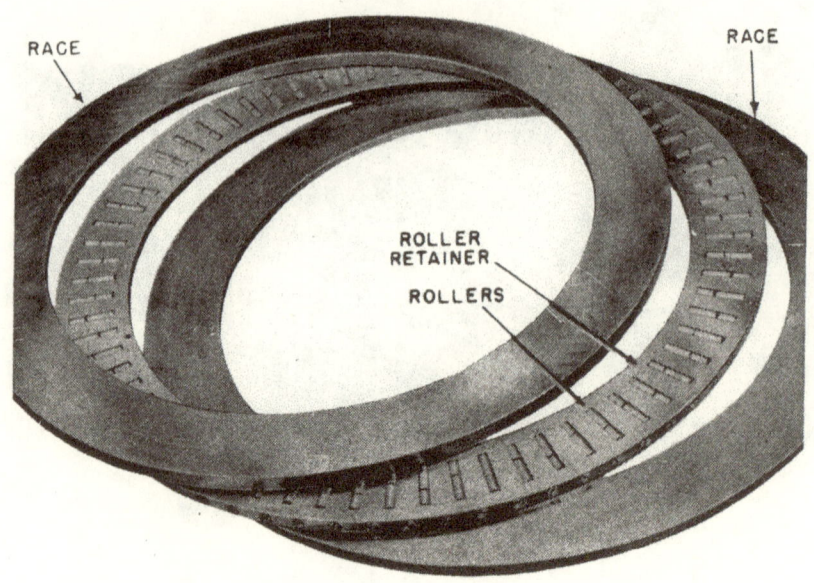

Figure 150. Roller bearing assembly, bottom carriage, 155-mm gun carriage M1.

Figure 151. Belleville springs in bottom carriage, 155-mm gun carriage M1.

the exception of the pinion, the gear train is mounted on antifriction bearings. The elevating mechanism is equipped with a brake which is engaged at all times. It must be released by pressing down on the brake lever before the gun is elevated or depressed. As soon as the required elevation is attained, the pressure on the brake lever is released, causing the brake to resume its normal position.

g. TRAVERSING MECHANISM. The traversing handwheel is connected through a flexible joint to a worm which drives a worm gear connected to the pinion. The pinion, in turn, meshes with the traversing rack on the bottom carriage. Like the elevating mechanism, the worm and worm gears are mounted in antifriction bearings.

h. TRAILS AND SPADES. Fabricated steel trails are attached by trail hinge pins to the bottom carriage. In firing, they prevent the movement of the carriage to the rear. For the opened position, each trail is moved about 30° from the coupled position and secured to the ground by heavy spades keyed to the trail ends (fig. 144). Front spades, attached to the bottom of the carriage, are also provided. They are assembled in the front spade brackets when in the firing position.

i. BOGIE. The bogie (fig. 152) supports the weight of the gun carriage only when it is in the traveling position. Two axles are provided, the straight axle and the arch axle. Torque rods are included to eliminate strain. Each axle is provided with two dual-tired wheels mounted in antifriction bearings. The bogie is also supplied with a bogie-lifting mechanism to raise the wheels off the ground when in the firing position.

j. BRAKES. Bendix-Westinghouse air brakes are provided for each of the wheels. Compressed air is obtained through two air lines connected to an air compressor unit in the prime mover. A mechanical brake system secured to each of the bogie axles is also supplied.

k. TRAVELING POSITION. When placed in traveling position (fig. 147), the gun is detached from the recoil cradle and drawn to the rear and locked. This movement of the gun from the battery position or vice versa is done by means of the prime mover (fig. 153).

l. FIRING PLATFORM. (1) As in the case of the Panama mount, developed for the 155-mm gun G.P.F., firing platforms to increase the traverse of the 155-mm gun M1 have been devised. Figures 154 and 155 show such a platform emplaced.

(2) A central cylindrical base ring, supported by timbers, is buried in the ground with only the upper part extending above the surface. The base ring encloses a base plate (not shown) and a dome assembly. The latter contains a socket into which the bolster plate ball is inserted and held in place by a retaining ring, thus

forming the pivot about which the gun carriage is traversed. The bolster plate itself is bolted to the gun carriage. To give support to the base ring, it is attached to four anchors (not shown) by tie rods extending in four directions. The tie rods and anchors are buried.

Figure 152. Bogie, 155-mm gun carriage M1, rear view.

(3) The trails of the gun carriage are supported by a steel rail and timber assembly. This assembly is made up of separate sections and when emplaced, forms a circle 37 feet in diameter. The circle is held in place by radial members attached to the base ring.

(4) A 7½-ton, 6 x 6, Mack prime mover can transport the disassembled platform and tow the gun carriage. Eight men, manhandling the component parts, can emplace the platform in approximately 2 to 5 hours, depending upon the type of soil in the area of the emplacement, and can assemble the gun to the platform in 1

Figure 153. 155-mm gun M1, retracting cable attached for pulling gun into firing position.

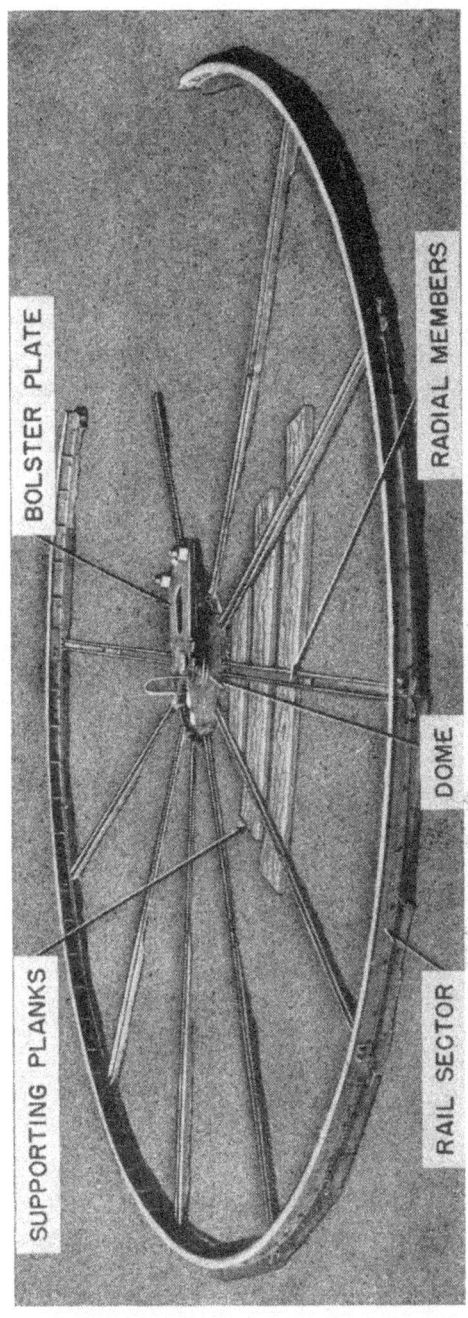

Figure 154. Firing platform for 155-mm gun carriage M1, emplaced, ready for emplacement of gun, rail sections not completely assembled.

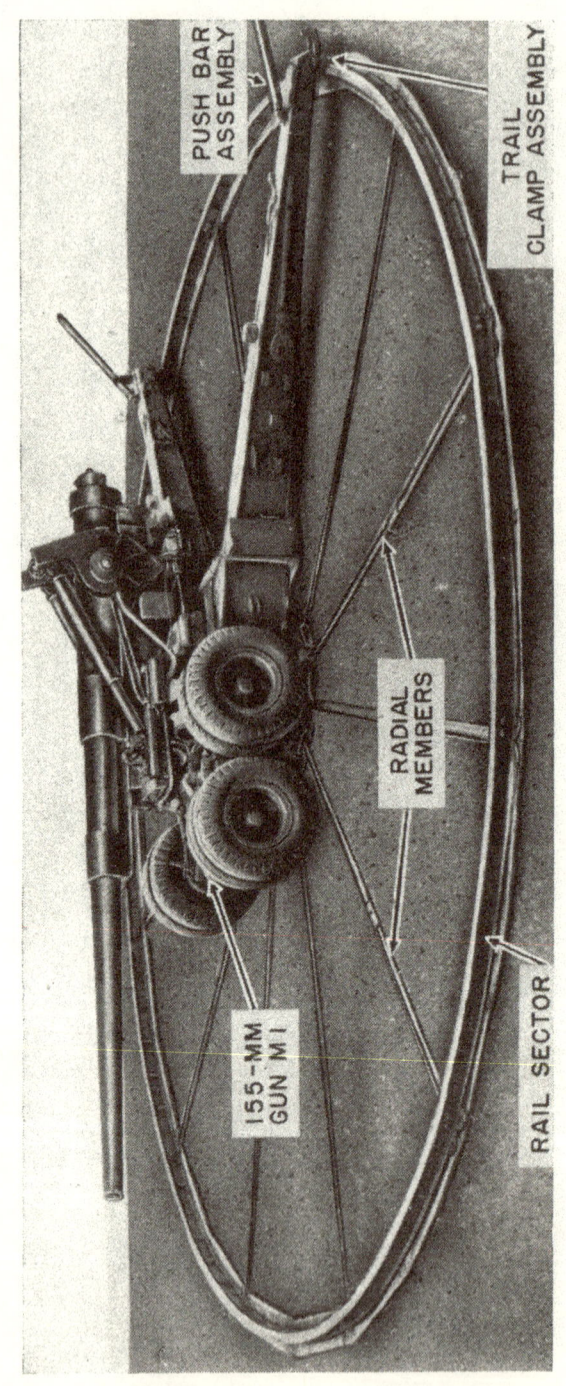

Figure 155. Firing platform with 155-mm gun carriage M1, emplaced.

hour. Working under the same conditions, the same number of men can disassemble the platform and stow it in a truck in 2 to 4 hours.

(5) The gun carriage can be traversed 360° without excessive effort by four men, two on each of the trail push bars. The gun can be fired with equal stability with the rails assembled for 180° or 360° traverse.

Section IV. RAILWAY ARTILLERY CARRIAGES

49. General

a. The earliest use of railway artillery by American armies was during the Civil War. Both Confederate and Union forces mounted guns and mortars on flat cars. Although essentially primitive in design, their tactical mobility was easily recognized. In the European War, 1914-1918, railway artillery was undeveloped during the initial phases of the conflict, but as the war progressed, both sides introduced the use of these heavy weapons to fire on targets such as communication systems far behind the lines. To meet this sudden demand, cannon from seacoast forts and old battleships were mounted on improvised railway cars and sent to the front. At first these mounts for major caliber cannon had no recoil or traversing mechanisms, but were allowed to roll along the tracks with the brakes set to absorb the shock of recoil. After each round, a winch was used to bring the car back to its original position. Traverse was secured by moving the mount along a curved track. Later developments in design included a means of pointing in direction in which the whole car was traversed on a previously prepared bed or emplacement.

b. The principal advantages of railway artillery as a seacoast defense weapon are threefold:

(1) Its mobility enables the Army to dispatch it to any coastal area in potential danger.

(2) Its tactical importance is paramount because its presence, even though known by the enemy, cannot be accurately ascertained since we possess numerous concealed positions to which railway mounts may be moved when deemed necessary.

(3) The tremendous fire power, such as that of the 8-inch gun, makes it a valuable asset to any defensive position.

50. 8-inch Gun Railway Mount M1A1

a. GENERAL. (1) The 8-inch gun Mk. VI Mod. 3A2 was originally designed for naval use. However, due to its flexibility of operation it has been accepted for Army use. Because of its mobility, un-

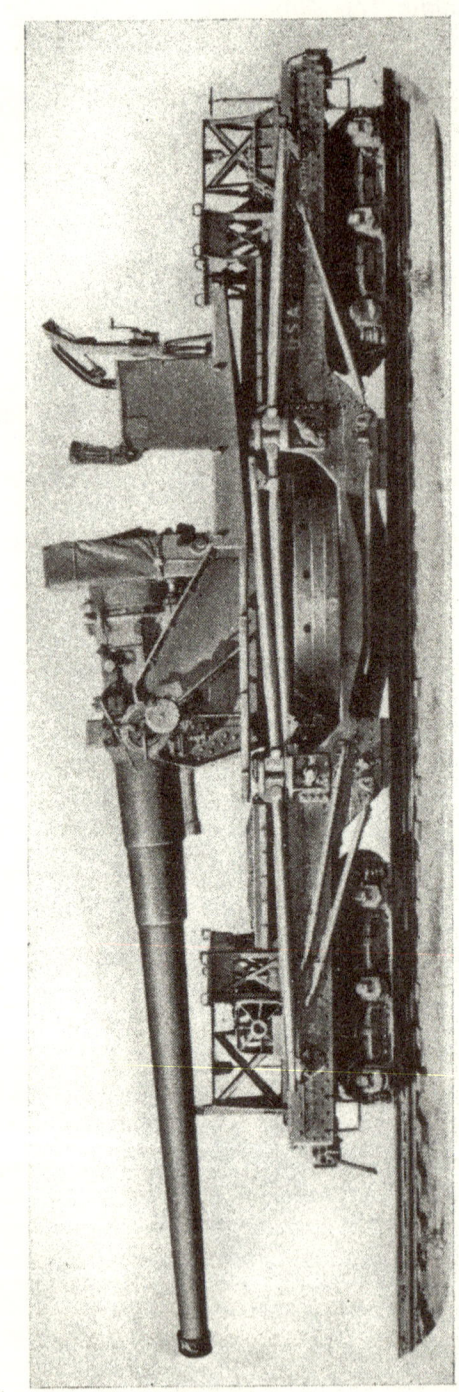

Figure 156. 8-inch gun Mk. VI Mod. 3A2 on 8-inch gun railway mount M1A1, traveling position.

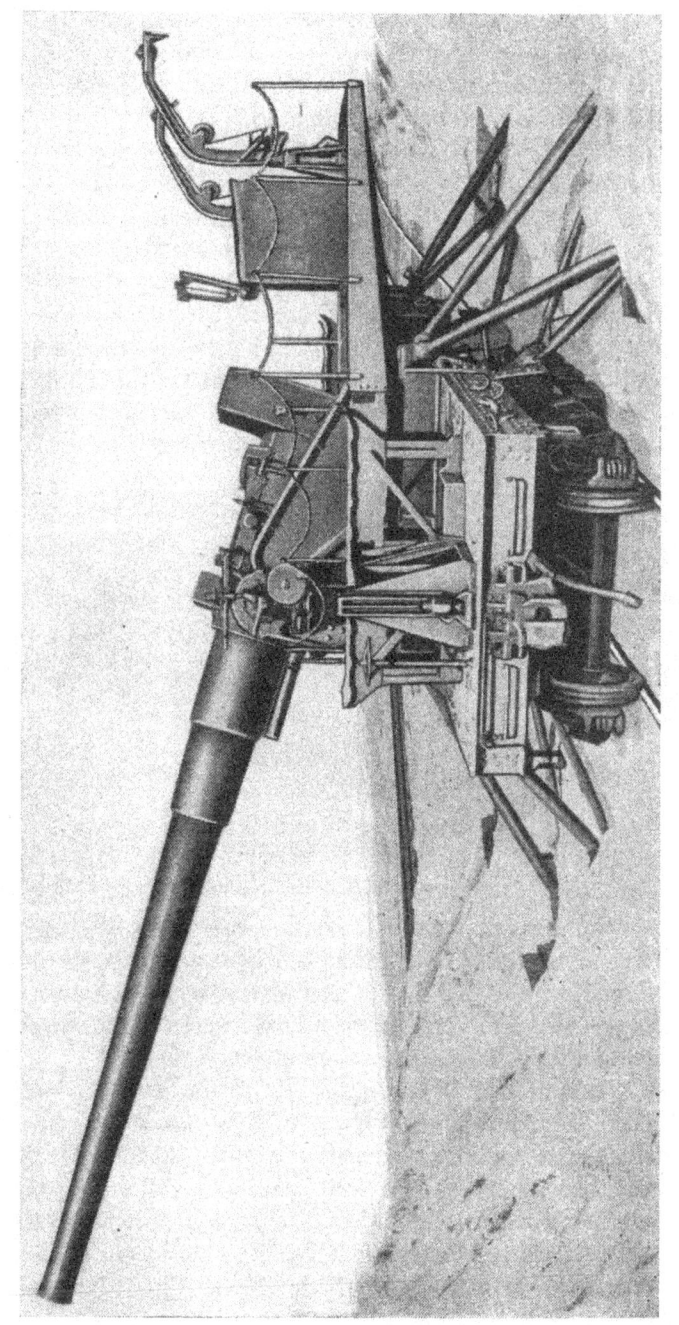

Figure 157. 8-inch gun Mk. VI Mod. 3A2 on 8-inch gun railway mount M1A1, firing position.

limited traverse, and effective range of over 35,000 yards, this 8-inch railway gun (fig. 156) is an outstanding mobile weapon. It is also used in fixed emplacements as described in paragraph 41.

(2) As may be seen from figure 157, the mount is practically a barbette carriage mounted upon a railway car. The elevating and traversing mechanisms, racer, side frames, platform, trunnions and trunnion bearings, cradle, and recoil system are essentially the same as those described in paragraph 41 for the same gun in a fixed emplacement. Also, as on the fixed gun, there is a brake to hold the tube at the desired elevation and a heavy counterweight at the breech end to enable the trunnions to be mounted well toward the rear, without the use of equilibrators. The base plate, which corresponds to the base ring on the fixed mount, is a one-piece steel casting used as the center section of the car proper, each end being bolted to the car frame. The upper part provides a bearing surface upon which conical rollers support the top carriage.

Figure 158. 6-wheel truck assembly, 8-inch gun railway mount M1A1.

b. CAR BODY. The car body is a drop-frame type, equipped with two traveling trucks (fig. 158). Each truck has six wheels and is of a built-up construction. The load is transmitted to the journals by semielliptic springs and equalizers. Each truck is equipped with two independent air brake systems of the latest design.

c. JACKS. This mount is designed for firing from a track emplacement (fig. 157), but it is not supported by the car trucks during firing. Instead, the mount is raised by hand-operated lifting jacks so that eight pedestal jacks (fig. 160) can be placed on the cross ties adjacent to the rails. The car is then lowered until it rests on these jacks. The trucks are not removed but do not support any of the weight. There are four built-in lifting jacks (fig. 159) located in the corners of the base plate directly above the rails of a standard gage railroad track. Each jack consists of a ram and bronze nut through which the lifting screw operates.

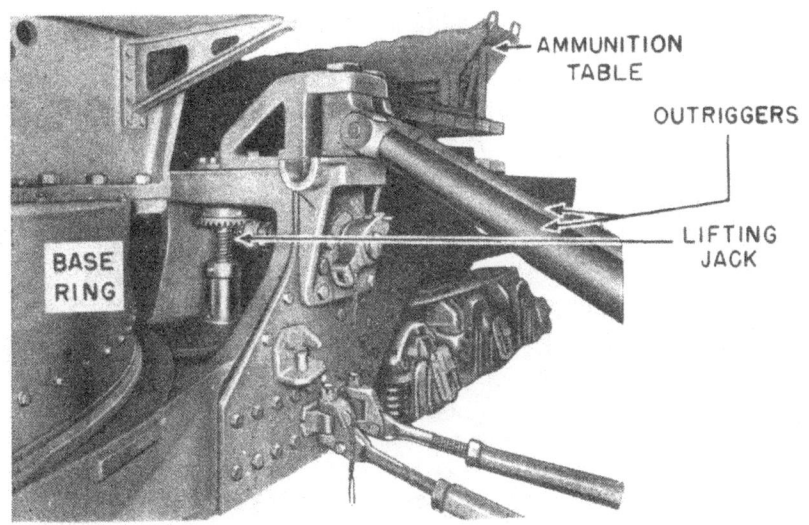

Figure 159. Lifting jack mechanism, 8-inch gun railway mount M1A1.

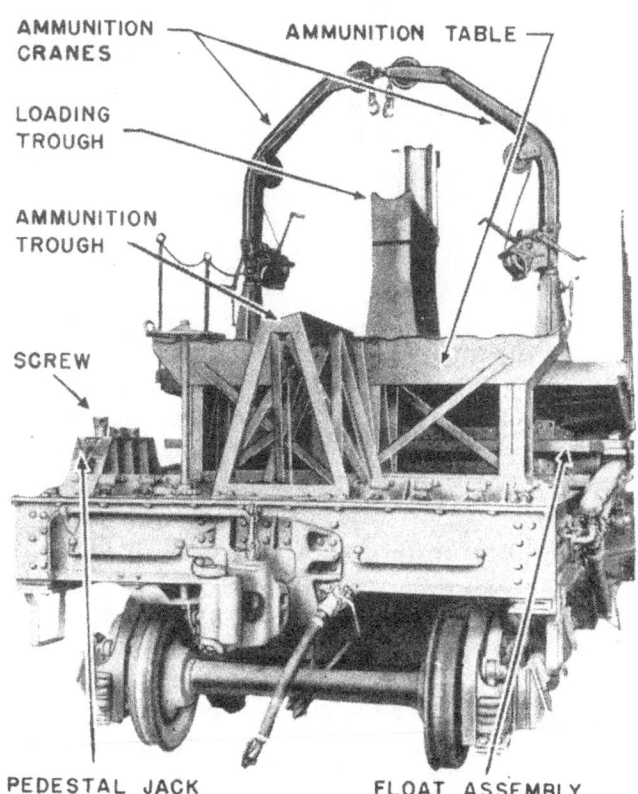

Figure 160. 8-inch gun railway mount M1A1, rear view.

161

d. OUTRIGGERS. Eight outriggers (fig. 157), four on each side, act as supports to prevent the mount from tipping or sliding when the gun is fired. They are made of steel tubing with adjusting screws in the lower ends. The adjusting screws have ball ends resting in sockets in the footplates. These footplates distribute the load over wooden floats.

e. AMMUNITION TROUGHS, TABLES, AND CRANES. Two ammunition troughs (fig. 160) of built-up structural steel are fastened to the floor of the car, one at each end of the car. They are used to transfer shells from an ammunition car to the ammunition tables. Each ammunition table has a capacity for nine shells. Ammunition is hoisted from the tables to the loading troughs by hand-operated loading cranes. These cranes are locked together when the mount is in the traveling position.

51. 8-inch Gun Railway Mount M1918

a. GENERAL. The 8-inch gun railway carriage M1918 (fig. 161) was developed during the European War, 1914-1918. The mount is similar to the 8-inch gun railway mount M1A1. It is of the center-pintle type, permitting the gun to be traversed through 360°. The gun is mounted in a cradle with one hydraulic recoil cylinder and four spring recuperator cylinders. The cradle is mounted on a barbette-type carriage. The base ring is firmly attached to the car body, and a working platform is built around the top carriage for the gun crew. Two crane masts are used for handling projectiles from the ground to the loading tray.

b. RECOIL AND COUNTERRECOIL SYSTEMS. The recoil is controlled by one conventional hydraulic cylinder mounted on the bottom of the cradle. Two of the spring recuperator cylinders are mounted above and two below the cradle. In order to withstand the stresses sustained in recoil, the length of recoil has been made long (48 inches). A dashpot counterrecoil buffer at the forward end of the recoil piston rod prevents the mount from charging into battery.

c. ELEVATING AND TRAVERSING MECHANISMS. This mount employs a conventional type of elevating mechanism which consists of a circular rack fastened to the cradle and a set of plain spur gears connected to the elevating handwheel on the top carriage. The traversing mechanism employed is the rack-and-pinion type.

d. AMMUNITION. The gun was originally designed for operation on a fixed barbette carriage, firing a heavy seacoast projectile. However, sustained operation of the railway model using a heavy projectile is possible only when firing parallel to the tracks, because at other azimuths the mount is not stable. At present these guns are fired with the lighter 200- and 260-pound projectiles.

Figure 161. 8-inch gun M1888 on railway mount M 1918, emplaced.

52. 14-inch Gun Railway Mount M1920

a. GENERAL. (1) As a result of the successful use of the Allied 14-inch railway batteries during the European War, 1914-1918, the 14-inch gun railway carriage M1920 was developed. Firing from a prepared fixed emplacement which permits a 360° traverse, maximum effectiveness of the weapon against primary targets is achieved. To attain maximum delivery of fire power, mobility is sacrificed to some extent. However, its value in harbor defense is considerable, as the 14-inch gun M1920MII is capable of firing a 1,560-pound, armor-piercing projectile at ranges exceeding 40,000 yards. The mount permits gun elevations from $-7°$ to $+50°$.

(2) The recoil mechanism is similar to that on the 16-inch gun barbette carriage M1919 and the 16-inch howitzer barbette carriage M1920 described in paragraphs 38 and 39. It consists of four conventional recoil cylinders and one pneumatic recuperator cylinder.

b. TOP CARRIAGE LIFTING MECHANISM. One of the problems ingeniously solved in the design of this gun was the requirement for two different heights of trunnions. To lower the center of gravity for traveling and to decrease the over-all height of the mount so that it would pass through tunnels, a low height of trunnions was desirable. On the other hand, the trunnions had to be high for firing so that the gun would have ample space for recoil at high elevations. These mutually exclusive requirements were met by the use of a movable top carriage which is raised for firing and lowered for traveling. The general idea is illustrated in figures 162 and 163 which show the top carriage in both upper (firing) and lower (traveling) positions. The carriage-lifting mechanism consists essentially of two lifting brackets with raising screws (fig. 164). The raising screws are capable of lifting approximately 280,000 pounds.

c. FIELD EMPLACEMENT. For land warfare it is desirable to fire the gun from a field emplacement on the track. This is known as a field emplacement. For this emplacement a previously constructed curved spur is used and emplacement can be accomplished in approximately 8 hours time. The mount is run to the proper position on the spur which will permit the gun to be directed at the target. Steel I-beams parallel to the tracks (fig. 164), which correspond to pedestal jacks (par. 50c), are spiked to the ties, and the mount is lowered and supported on the I-beams. Additional support in firing is afforded by the rear trucks and by six outriggers which extend from the sides of the mount to steel floats embedded in the ground. In this emplacement, the gun has a total traverse of 7°. Traverse is accomplished by hand operation of the top carriage traversing mechanism which pivots the top carriage at its rear end or upper pintle (fig. 164).

d. FIXED EMPLACEMENT. (1) For seacoast defense work, which requires a much greater traverse than 7°, it is necessary to have a previously prepared platform or permanent emplacement (fig. 165). This emplacement, providing a traverse of 360°, consists of a concrete block about 42 feet in diameter. To this block are bolted two steel parts—the circular base plate, about 9 feet in diameter, and the base ring, about 32 feet in diameter (fig. 166). To go into position on this emplacement, the mount is moved over the bridge rails and hinge rails until the lower pintle center coincides with the center of the concrete block. The whole mount is then lowered by a mechanical lowering device until its weight rests on the base plate. The base of the lower pintle is then bolted solidly to the base plate. The two traversing rollers on the mount traversing beam rest on the base ring. Clips (fig. 167) engage the lip on the edge of the base ring and prevent the mount from tipping. The top carriage is then raised to firing position, and the trucks are run from under the mount. As the electric power plant is carried on the forward truck, this truck may be moved only to the length of the cable furnished, about 100 yards. The total time required for emplacing the mount after the concrete block is completed is about 2½ hours.

(2) For long-range firing and particularly for augmenting the fire power of seacoast fortifications, this gun may be dispatched to any one of a number of fixed emplacements along the coast and be ready to fire within a short period of time.

e. MOUNT-TRAVERSING MECHANISM. The mount-traversing mechanism is employed when the mount is on the permanent emplacement. As shown in figures 165, 166, and 167, the mechanism consists of a pedestal (or pivot), around which the mount revolves, and a traversing beam and traversing roller assembly. When the pedestal is in position on the base plate, the traversing rollers rest on the base ring. As the center of gravity of the mount is between the pedestal and the traversing beam, part of the weight is borne by each of these members. Actual traverse is accomplished by turning the traversing rollers through suitable gearing. During recoil, the stresses imposed on the mount-traversing beam are very great. Therefore, the traversing mechanism is equipped with an antifriction device which will permit easy traversing and at the same time enable the beam to withstand these stresses. The two traversing rollers, which are attached to the traversing beam, are supported by Belleville springs. When these rollers are subjected to the firing load, the Belleville springs are compressed and the traversing beam itself rests on the base ring, directly transmitting the stress of recoil to the base ring. The mount may be traversed by hand or by power.

f. ELEVATING MECHANISM. The elevating mechanism is of conventional design, employing a circular rack and spur gearing. For

Figure 162. 14-inch gun on railway mount M1920, firing position, field emplacement.

Figure 163. 14-inch gun on railway mount M1920, traveling position.

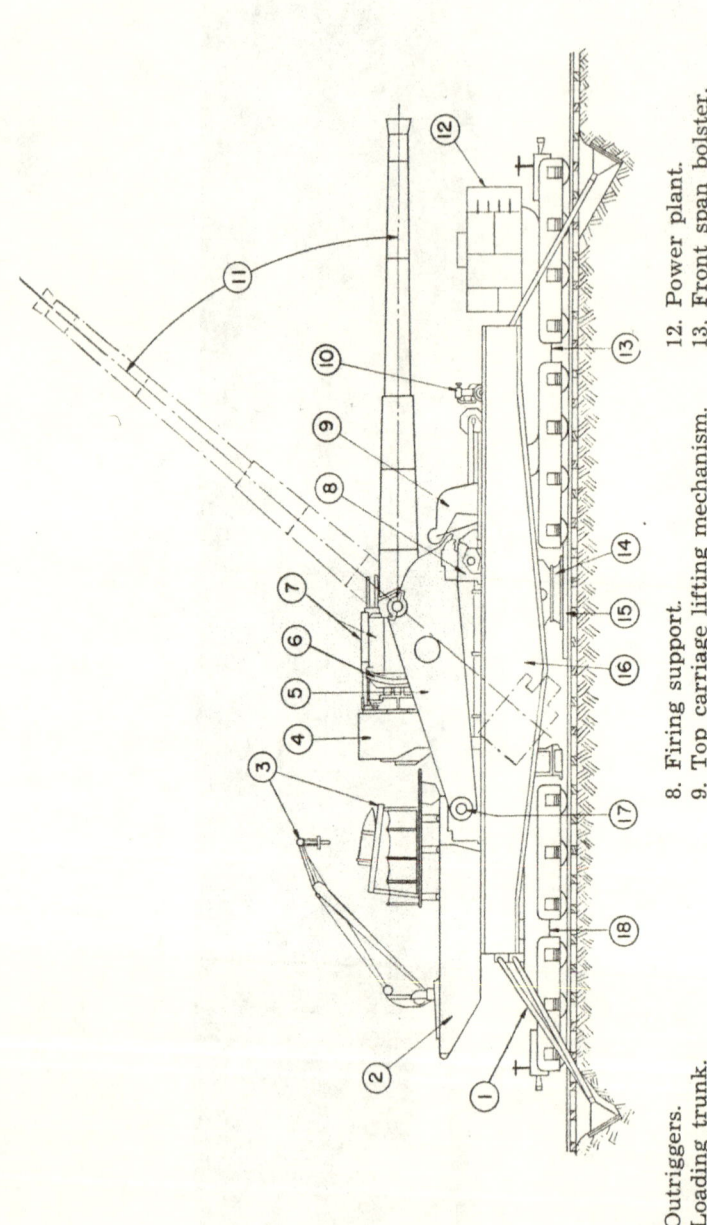

1. Outriggers.
2. Loading trunk.
3. Loading mechanism.
4. Recoil band.
5. Side frame (top carriage).
6. Elevating rack.
7. Recoil and recuperator cylinders.
8. Firing support.
9. Top carriage lifting mechanism.
10. Air compressor.
11. Maximum elevation 50°; maximum elevation at which gun can be fired from field (track) emplacement is 19°.
12. Power plant.
13. Front span bolster.
14. Lower pintle.
15. I-beam.
16. Car body girder.
17. Upper pintle (universal joint).
18. Rear span bolster.

Figure 164. 14-inch gun on railway mount M1920, firing position, field emplacement.

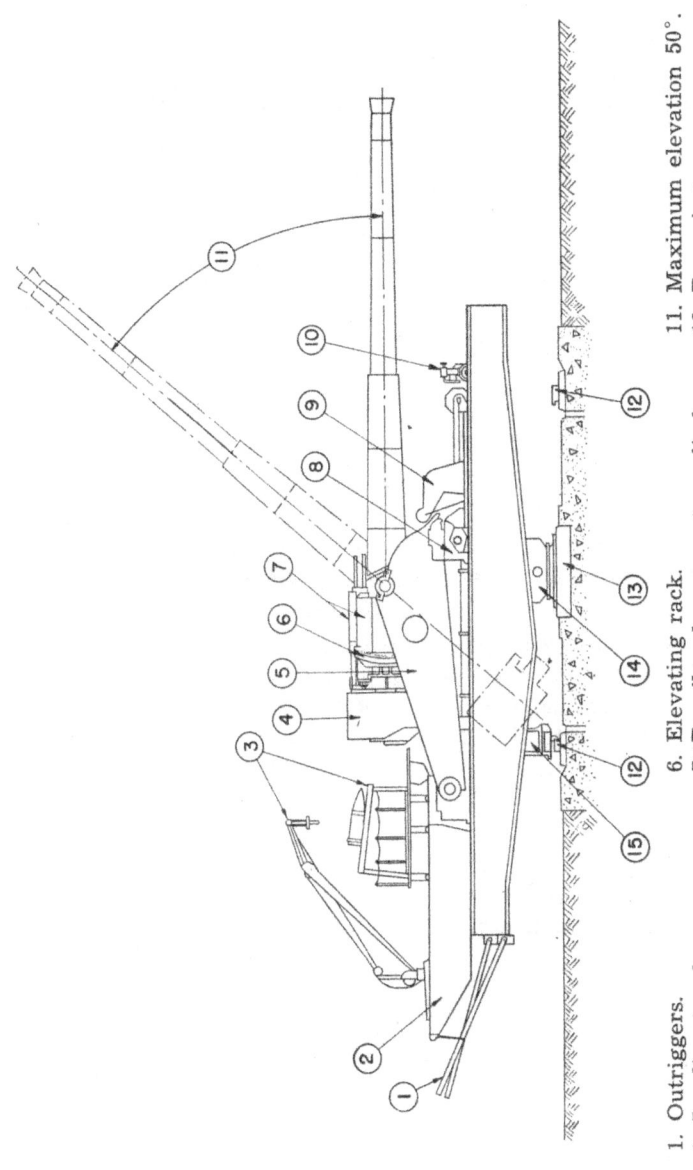

1. Outriggers.
2. Loading trunk.
3. Loading mechanism.
4. Recoil band.
5. Side frame (top carriage).
6. Elevating rack.
7. Recoil and recuperator cylinders.
8. Firing support.
9. Top carriage lifting mechanism.
10. Air compressor.
11. Maximum elevation 50°.
12. Base ring.
13. Base plate.
14. Lower pintle assembly.
15. Mount traversing beam.

Figure 165. 14-inch gun on railway mount M1920, fixed emplacement.

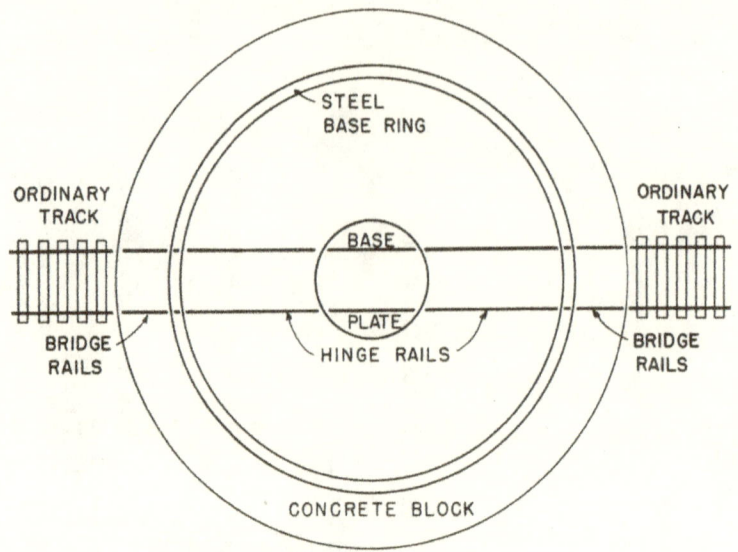

Figure 166. Diagram of fixed emplacement, 14-inch gun railway mount M1920.

elevating and depressing by electric power, the mount is furnished with a self-contained power plant mounted on the front truck.

g. LOADING MECHANISM. Mounted on the loading platform are the loading cranes, loading tray, and spanning tray. There are two cranes, one for powder and one for projectiles. For loading, the gun is depressed to $-7°$ and the projectile rammed by sliding it down the tray, which is also set at a $-7°$ slope. A slight manual impetus will start the projectile down the tray and, with its velocity increased by gravity, embed the rotating bands firmly in the rifling of the gun.

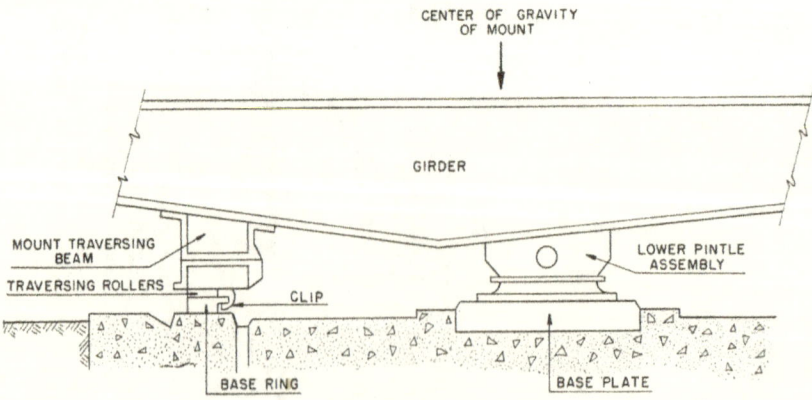

Figure 167. Diagram of traversing mechanism, 14-inch gun railway mount M1920, fixed emplacement.

53. Ammunition and Auxiliary Cars

Railway batteries are supplied with several types of standard cars such as ammunition, store (box), and fire control. These cars differ very slightly from the commercial cars of the same type. The fire control car (fig. 168) is provided with windows and heating arrangements. The ammunition car has special ends. Each end consists of a folding-back floor plate and an end-swing door that permits the service of projectiles and powder from the ammunition car to the mount car. This is accomplished by means of an overhead rail trolley that can be slid forward and projected over the end of the mount car. On this trolley are mounted ordinary rolling triplex blocks. In the loading of the cars a limit is placed upon the amount of ammunition per car. In the case of certain guns, complete rounds (powder and projectile) are carried in the same car; in others, the projectile and powder are carried in separate cars.

Figure 168. Fire control car, railway artillery.

CHAPTER 5

SEARCHLIGHTS

54. General

a. All U.S. Army searchlights depend on the electric arc (fig. 169) as their source of light. The arc is formed between two carbons (fig. 171), one known as the negative and the other as the positive. Because the core of the positive carbon on the modern searchlight is composed of rare earths, cerium and lanthanum, an arc of much

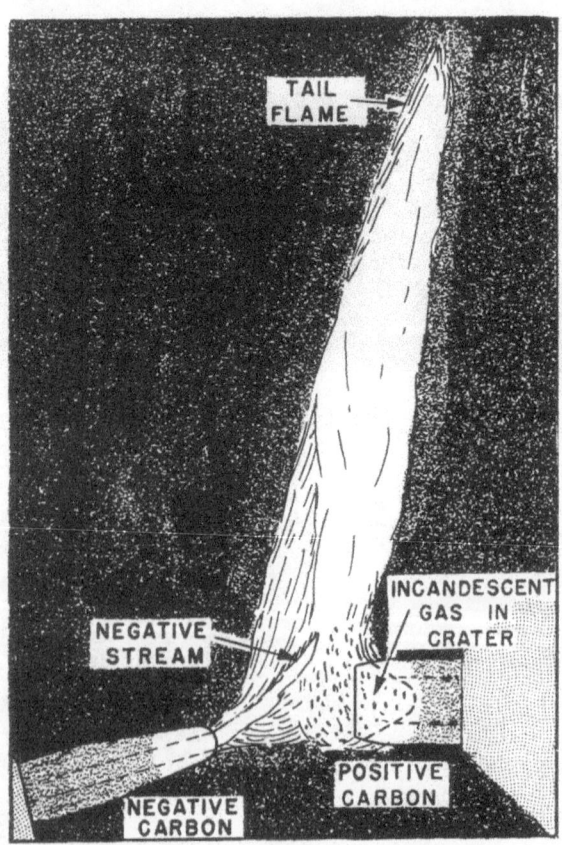

Figure 169. Drawing showing normal arc of searchlight.

greater intensity is obtained than with the ordinary carbon arc employed in older searchlights. The actual light source is an incandescent ball (globule) of vapor, formed in the high-intensity arc and held in a crater in the positive carbon. This globule of incandescent gas may be thought of as a miniature sun.

  b. The passage of the current causes the stream of positive ions to vaporize and volatilize the materials of the positive carbon. However, the volatilization of the core of the carbon takes place at a faster rate than that of the solid carbon shell, thus forming a crater filled with incandescent gas. The positive carbon must be continuously rotated to maintain a crater wall of uniform height. A reflecting parabolic mirror is located in the base of the searchlight drum. The globule of gas must be kept at the exact focus of this mirror, as tests show that a 40-percent loss of efficiency occurs if the light is ⅛ inch out of focus. Therefore, in addition to the rotation of the carbon, means must be provided for gradually feeding both positive and negative carbons to compensate for their burning and to maintain the globule of gas at the exact focus of the mirror. In early searchlights this rotating of the positive carbon and feeding of both carbons were done by hand. Experience showed that it was difficult for an operator to accomplish this manually; therefore, mechanical means were provided.

  c. The mirror, being parabolic in shape, reflects the light rays forward in an almost parallel pattern, producing a well-defined narrow beam of light. The primary purpose of this beam of light is to illuminate hostile naval targets so that fire may be quickly and effectively directed upon them. To fulfill this purpose, the location and effective range of the searchlight are of paramount importance. The range of a searchlight is the maximum distance at which an object in its beam is visible. The visibility of an object, however, does not depend on its actual illumination but upon the contrast between its illumination and the illumination of the surrounding field. The most obvious way of improving the contrast between the illuminated target and the searchlight beam is to move the controlling observer away from the searchlight. This decreases the depth of illuminated atmospheric particles through which he sees the target. The luminosity of the field is thereby decreased, improving the contrast and increasing the range. Experience has shown that the distance between the controller and searchlight should be at least 50 feet. Each increase in distance up to that point brings about a substantial increase in range. Distant control also enables the light to be controlled directly from the observation post of the commander of the particular tactical unit using the light. All existing 60-inch searchlights are supplied or may be supplied with modification devices for spreading the beam. Such modifications provide additional means of tactical employment because it is possible to spread the narrow beam of about 1¼° to a maximum spread of about 15°. However, as the beam is spread, the range of illumination is rapidly reduced.

d. The rotation of the positive carbon is a simple motion accomplished through gearing (16, fig. 170) by an electric motor. The feeding of the positive carbon is a much more complicated problem as the carbon must be fed at a variable rate, depending on how fast it is being consumed. In modern harbor defense searchlights such as the Sperry, use is made of the heat rays from the positive carbon playing upon a thermostat (9, fig. 170) to control this motion. The thermostat is located inside a small window. A focusing lens (mirror on some types) is employed to focus the rays from the arc on or near this window. When the light source is in exact focus, the heat rays are directed just outside the window. As the carbon burns away, the light source leaves the focal point and the beam enters the window, causing the thermostat to heat. This closes an electrical contact which in turn causes an electromagnet (11, fig. 170) to feed the positive carbon forward by means of the ratchet and positive feed gear assembly (10, fig. 170). This movement of the positive carbon brings the light source back into focus and causes the rays from the lens to leave the window. The thermostat then cools, and the feeding of the positive carbon stops. When the positive carbon again burns away, the rays enter the thermostat once more and the feeding operation is repeated.

e. In addition to maintaining the light source at the focal point of the mirror, it is necessary to maintain the arc at proper length. This is accomplished by regulating the feeding of the negative carbon. One feeding device is designed to maintain a constant voltage across the arc. To accomplish this, a coil (21, fig. 170) is connected across the arc. The reciprocating shaft (15, fig. 170) and the negative feed pawls (3, fig. 170) attached to it are moved right and left by the rotary motion of the eccentric disk (14, fig. 170) rotated by the feed motor. When the arc becomes too long and the voltage across the coil too high, it attracts an armature (19, fig. 170) which raises the negative feed pawls, engages the lower pawl, and feeds the negative carbon forward by rotating the negative control rod (2, fig. 170). If the arc becomes too short, the attraction of the coil for the armature is lessened and a coil spring (20, fig. 170) pulls the armature back and lowers the negative feed pawls, thus engaging the upper pawl and causing the negative carbon to be retracted.

f. Both the negative and the positive carbon feeding operations go on constantly and automatically. The light source is thus main-

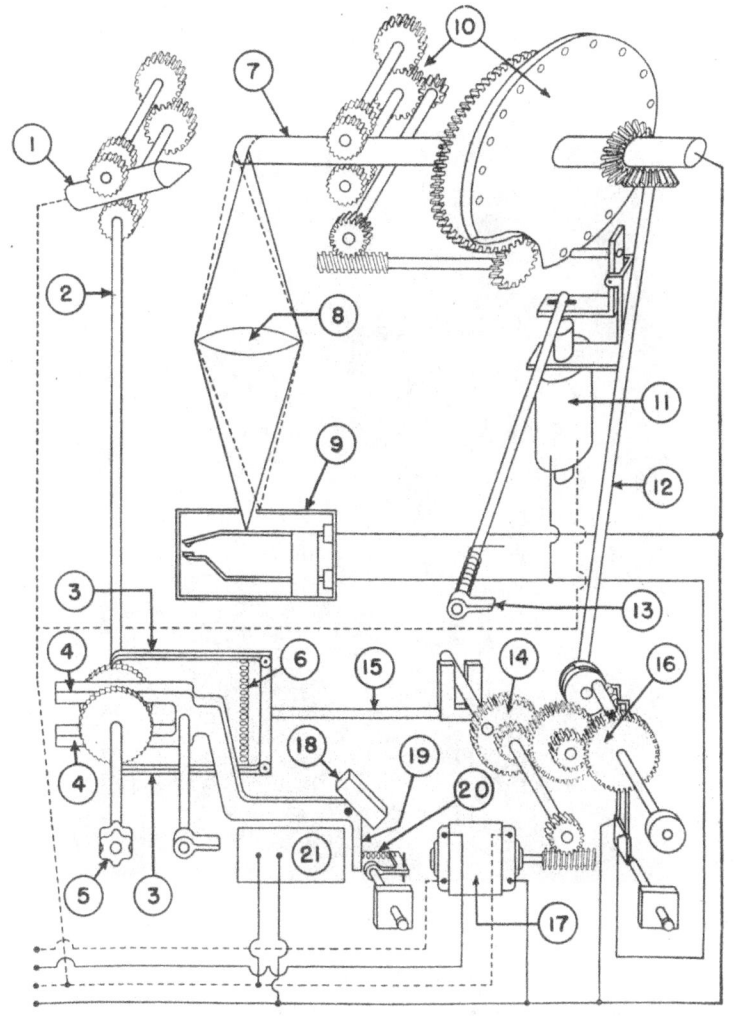

1. Negative carbon.
2. Negative control rod.
3. Negative feed pawls.
4. Pawl guards controlling negative feed.
5. Negative hand feed knob.
6. Negative feed pawl spring.
7. Positive carbon.
8. Thermostat lens.
9. Thermostat.
10. Ratchet and positive feed gear assembly.
11. Positive feed control electromagnet.
12. Positive control rod.
13. Positive hand feed lever.
14. Eccentric disk and gear.
15. Reciprocating shaft.
16. Gear train.
17. Feed motor.
18. Counterbalance.
19. Balanced armature.
20. Adjustment spring.
21. Arc length control coil.

Figure 170. Lamp and control mechanism, Sperry searchlight, schematic diagram.

tained at the focal point of the mirror and the arc length is kept constant. The negative feed mechanism also serves to strike the arc when the main switch is closed.

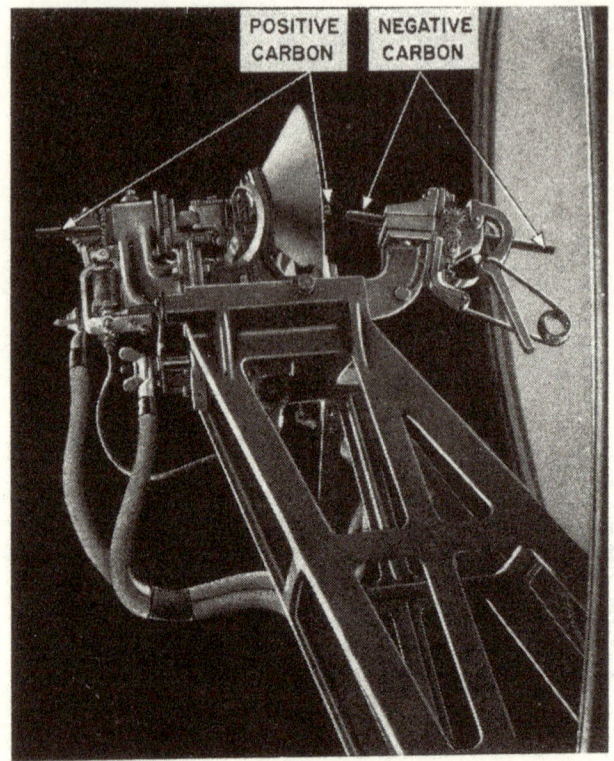

Figure 171. Lamp unit, Sperry searchlight.

55. Fixed Harbor Defense Searchlights

Harbor defense searchlights are of two types, fixed and mobile. The fixed type is being replaced by the mobile light.

56. Mobile Searchlights

a. GENERAL. A majority of the mobile seacoast artillery searchlight units now being used are either of General Electric or Sperry manufacture. Both of these searchlights are manufactured under the same specifications and their basic operating principles are similar. The differences are in nomenclature and in the methods which each company has evolved to satisfy the specifications according to its own designs and manufacturing facilities.

b. SEARCHLIGHT M1941. (1) The source of illumination in the Sperry or the General Electric searchlight M1941 (figs. 172 and 173) is a direct-current electric arc of 150 amperes at 78 volts. The searchlights use high-intensity carbons which create an extremely hot arc containing an incandescent globule in the crater of the positive carbon. The illumination from the arc is automatically maintained at the focal center of a 60-inch parabolic metal mirror with a total beam spread of $1\frac{1}{4}°$ when the searchlight is in focus.

(2) The lamp is a unit (fig. 171) which holds the carbons and is housed in the drum. The lamp control mechanism is located on the outside of the drum. As explained in paragraphs 54d and 54e, the function of this mechanism is to operate the arc at maximum efficiency by moving the carbons. Electromagnetic action controlled by a thermostat provides automatic feeding of the positive carbon. The electromagnetic force of a coil pulling in opposition to an adjustment spring provides negative carbon control (arc length control). Variations in the arc length cause variations in the arc voltage and the arc current. In a Sperry searchlight, the pull of the electromagnetic coil is regulated by the voltage which varies directly with the arc length. In a General Electric searchlight, regulation is achieved by the current which varies inversely with the arc length. In either case, the negative carbon is moved backward or forward by the action of the electromagnetic coil and adjustment spring to maintain the proper arc length. In case of failure of the automatic control, the movement of the positive carbon may be regulated by a semiautomatic arrangement or by manual control. A ground-glass finder is used to indicate visually the position of the positive carbon in semiautomatic and manual control. Negative carbon control may also be maintained by hand if necessary.

(3) A ventilating fan mounted on top of the searchlight cools and sweeps the interior of the drum free of hot gases. Blue glass arc-view "peep sights" on either side of the drum are used for observing the arc. An arc switch is used to open and close the arc circuit. The Sperry searchlight has incorporated in this arc switch a thermal-operated, time-delay circuit breaker which provides for an instantaneous arc strike. The searchlight is moved in azimuth and elevation by means of a distant electric control (D.E.C.) from the control station. However, it may also be moved manually by the use of an extended hand controller.

c. CONTROL STATIONS M1941. The Sperry or General Electric control station (fig. 174) consists of a control unit and binocular mount which are mounted on a tripod. The control station is connected to the searchlight by a 500-foot, 15-conductor cable. The control unit provides for the remote control of the searchlight in azimuth and in elevation. The remote-control mechanism is driven by two sets of handwheels, one for azimuth and one for elevation.

1. Lamp control mechanism box.
2. Elevation gear sector.
3. Elevation zero reader.
4. Azimuth zero reader.
5. Hand controller socket.
6. Elevation control motor.
7. Ventilating fan motor.
8. Front drum.
9. Lamp unit.
10. Arc view peep sight.
11. Ground-glass finder.
12. Azimuth data receiver housing.
13. Mirror.

Figure 172. Sperry searchlight M1941, front quarter view.

1. Lamp control mechanism box.
2. Elevation zero reader.
3. Azimuth zero reader.
4. Hand controller socket.
5. Arc and elevation control box.
6. Ventilating fan motor.
7. Lamp unit.
8. Arc view peep sight.
9. Ground-glass finder.
10. Front drum.
11. Mirror.
12. Positive carbon.
13. Azimuth control box.

Figure 173. General Electric searchlight M1941, front quarter view.

The control unit also controls the movement of the observer's binoculars, which may be made to search the area about a moving point in space as established by the detector data. Zero readers indicate to the control station operators the direction in which the searchlight should be moved in order to keep it pointing in the direction determined by the detector.

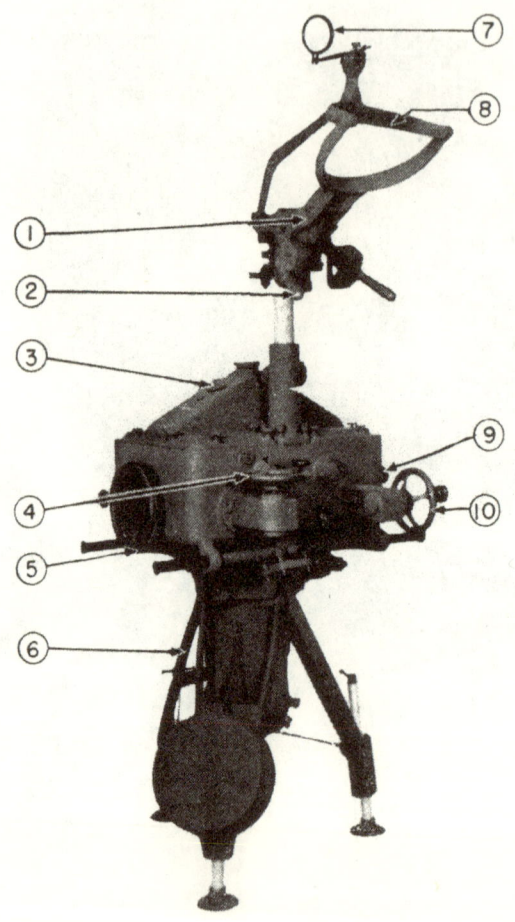

1. Binocular elevation zero marker.
2. Binocular mount adjustment handles.
3. Azimuth zero reader.
4. Observer's azimuth handwheel.
5. Azimuth zero reader handwheel.
6. Tripod.
7. Open sight.
8. Binocular mount.
9. Elevation zero reader handwheel.
10. Observer's elevation handwheel.

Figure 174. Sperry control station M1941.

d. DISTANT ELECTRIC CONTROL SYSTEM. The distant electric control (D.E.C.) system provides the means whereby the movements of the searchlight can be controlled electrically from the control station. The Sperry system consists of a pair of direct-current, step-by-step transmitters mounted in the control station and a pair of motors mounted on the searchlight. One motor drives the searchlight in azimuth and the second motor drives it in elevation. Both motors are controlled by their respective transmitters at the control station.

e. POWER PLANTS M1941. The electrical energy for the Sperry or General Electric searchlight M1941 is generated by a mobile power plant consisting of a gasoline engine which drives a direct-current generator. Power from this plant is used in the operation of the searchlight and the distant electric control system. The power plant is connected to the searchlight by two 200-foot, single-conductor cables, one positive and one negative.

f. SEARCHLIGHT M1942. The only fundamental difference between the Sperry and General Electric searchlights M1942 and earlier models is the method of distant electric control. The details involved are not within the scope of this manual.

g. TRANSPORTATION. Searchlight equipment is transported by two standard 2½-ton trucks. One truck tows an M1 searchlight trailer in which the searchlight is loaded, and the other tows the power plant (fig. 175). The automotive feature of the trucks, trailer, and power plant are covered in the maintenance manual for each vehicle.

Figure 175. Transportation of searchlight squad showing two 2½-ton trucks, M1 trailer, and searchlight power plant.

h. SPREAD BEAM LENS KIT. (1) For the purpose of illuminating motor torpedo boat targets at comparatively short ranges, a spread beam lens kit for the 60-inch portable searchlight has been provided. These lenses may also be used on searchlights for rapid searching of close-in water areas and illumination of beach defense areas and low altitude air targets.

(2) The kit includes all parts necessary for attaching four 24-inch molded glass lenses to the front door rim of the searchlight and a counterweight assembly to fit the back of the searchlight (fig. 176). The lenses are of a type commonly used for airport illumination and

Figure 176. Searchlight equipped with spread beam lenses. (Lenses in position for spread beam.)

are characterized by a series of uniform corrugations parallel to one axis of the plane of the lens. The holding ring for each lens consists of a narrow, one-piece, circular metal band shaped to fit the edge of the lens.

(3) Spread beam lenses can be attached in the field to Sperry searchlights, model 1934 and later, and to General Electric searchlights, model 1940 and later. The modification does not interfere with the long range of the normal beam of the light. In normal operation, the lenses lock in position for either normal parallel or fan-shaped spread beam illumination.

(4) Placing the four spread beam lenses perpendicular to the beam axis and directly in front of the source of the normal beam

results in spreading a considerable portion of the beam to a wider angle. When the four spreading lenses are positioned with the lens corrugations alined vertically, the total resultant spread in the horizontal plane is approximately 10°. The vertical spread remains approximately 1¼°. A choice of the number and combinations of the lenses to be used is possible. Thus, placing fewer than four lenses in the spread beam position reduces the intensity of illumination in the 10° sector and allows more of the light to travel out along the narrow beam. Such combinations as the upper two, lower two, and diagonally opposite lenses may be used if desired.

CHAPTER 6

DEMOLITION OF MATERIEL

57. General
Tactical situations may arise when lack of time and transportation will make it impossible to evacuate seacoast materiel. In such situations it is imperative that all materiel which cannot be evacuated be destroyed to prevent its capture intact by the enemy and its use against our own or allied forces. However, such destruction of artillery materiel should be carried out only as ordered by sector or higher command. The working principles to be followed are as follows:

a. The methods used must be adequate, uniform, and easily followed in the field.

b. If thorough destruction of all parts cannot be completed, the most important parts should be destroyed or removed. The same essential parts must be destroyed or removed on all like units. Such a procedure will prevent the enemy from constructing one or more complete units from the damaged ones.

c. All echelons should prepare plans for the destruction of materiel in the event of imminent capture. Such plans must be flexible enough to make allowance for variations in available time, equipment, and personnel.

d. All echelons should be trained to effect the desired destruction of materiel issued to them. However, training should not involve the actual destruction of materiel.

e. Destruction should be accomplished in the following priority:
 (1) Cannon (tube, breechblock, and recoil mechanism).
 (2) Carriage; railway artillery cars and vehicles.
 (3) Pneumatic tires.
 (4) Optical instruments.
 (5) Power equipment.

58. Cannon
The selection of a method of demolition will depend upon the tactical situation and the materials available. The methods for the destruction of the cannon are presented in the order of their effectiveness and desirability.

a. By HE Shell and Antitank Grenade or Antitank Rocket. Demolition of the cannon may be accomplished by using an unfuzed HE shell with either an antitank grenade or an antitank rocket as follows:

(1) Remove the recoil cylinder plug to allow the recoil fluid to drain. It is not necessary to wait until the cylinder is completely drained before proceeding with the next steps.

(2) Place an armed antitank grenade or antitank rocket in the bore, with the ogive nose toward the rear, approximately 6 inches in front of the HE shell to follow.

(3) Insert an unfuzed HE complete round into the cannon and close the breech. A base-detonating HE shell cannot be used in this method.

(4) Fire the cannon electrically if possible; if not, use a lanyard at least 100 feet long. The person firing the cannon should be under cover to the rear of the piece and approximately 20° off the line of fire. The danger zone is about 200 yards for 3-inch and 90-mm guns and about 500 yards for larger caliber cannon.

b. By HE Shells. Three-inch and 75-mm cannon may be destroyed by using HE shells as follows:

(1) Same as *a* (1) preceding.

(2) Jam an HE shell with point-detonating fuze in the muzzle of the cannon.

(3) Load a similar round in the cannon and close the breech.

(4) Fire the cannon as in *a* (4) above.

c. By TNT Blocks; TNT Blocks and HE Shells. (1) *57-mm and larger cannon.* Demolition of 57-mm and larger cannon may be accomplished with TNT blocks as follows:

(*a*) See *a* (1) preceding for first step.

(*b*) Insert TNT blocks near the muzzle and in the chamber of the cannon as indicated in (4) following. Close the breechblock as far as possible without damaging the safety fuse. Plug the muzzle tightly with earth to a distance of three times the caliber of the cannon. If it is not possible to plug the bore, a larger number of TNT blocks will be needed for effective demolition. Detonate the TNT charges simultaneously by means of a detonating cord (see (3) following).

(2) *155-mm to 16-inch cannon.* Demolition of 155-mm to 16-inch cannon may be accomplished with TNT blocks and HE shells in the following manner:

(*a*) See *a* (1) preceding for first step.

(*b*) Ram an HE shell (without base fuze) into the forcing cone and place TNT blocks behind it as specified in (4) following. Close the breechblock and detonate the TNT charge. The fuse may be routed through the primer hole in the spindle.

(3) *Danger zone.* The danger zone for both methods is from 200 to 500 yards. The person firing the charge should be under cover. Instructions on the wiring and firing of TNT can be found in FM 5-25.

(4) *Demolition charges.* The following number of ½-pound TNT blocks (with cardboard cases removed) will be needed with each cannon for effective demolition:

Cannon	Bore No. of blocks	Chamber No. of blocks
3-inch	2 to 3	4 to 6
90-mm	3 to 5	8 to 10
155-mm to 8-inch		30 to 50
12-inch		120
14-inch to 16-inch		150

d. By Incendiary Grenades. If evacuation is imminent and it is desired to accomplish demolition without telltale explosions, incendiary grenades should be used.

(1) *Procedure.* To accomplish demolition with incendiary grenades proceed as follows:

(*a*) Place unfuzed M14 incendiary grenades in the chamber. They should be placed on their sides and stacked one on top of another.

(*b*) Close the breech.

(*c*) Equip another incendiary grenade with a 15-second Bickford fuze, ignite it, and throw it in the muzzle.

(*d*) Elevate the cannon quickly to its maximum elevation. The metal from the grenades will fuse with that of the breech recess and breechblock making it impossible to open the latter.

(2) *Demolition charges.* The number of grenades to be used in the chamber per cannon is as follows:

Cannon	No. of grenades
6-inch (and smaller)	6 to 8
8-inch	10 to 12
12-inch to 16-inch	15 to 25

e. By Firing Cannon at Each Other. The least desirable method of demolition is by the following procedure: Fire adjacent cannon at each other at point blank range, using HE or AP shells. Fire from cover. The danger space is from 200 to 500 yards. The last cannon and carriage will have to be destroyed by the best means available.

59. Carriages

a. Fixed and Railway. Barbette carriages (fixed or railway) ordinarily will be adequately destroyed if the recoil cylinders are completely drained, and/or the recoil rods are disconnected from the recoil band when the cannon is fired for destruction as described in paragraph 58.

b. 155-MM GUNS. 155-mm gun carriages may be destroyed by detonating unfuzed HE shells with TNT blocks. One shell should be placed on each side at the junction of the trails and bottom carriage. The shells should be placed upright and a ½-pound TNT block put over the booster in each. Detonation should be simultaneous with that in the tube (par. 58). Another method which may be used is to detonate thirty ½-pound TNT blocks placed on the top carriage under the cradle in the vicinity of the pintle.

60. Pneumatic Tires

Even if time does not permit the destruction of the remainder of the vehicle, destroy tires by one of the following methods:

a. When combined with the destruction of the vehicle or artillery carriage by TNT, ignite an incendiary grenade under each tire. Be certain fires are well started before detonating TNT.

b. Damage the tires with an ax, pick, or heavy mechine-gun fire; pour on gasoline and burn.

61. Breechblock

If the breechblock is not destroyed by any of the above methods, it should be rendered useless with a heavy sledge.

62. Recoil Mechanism

The recoil mechanism will be effectively detroyed by the methods described in paragraph 58.

63. Power Equipment

All power equipment should be rendered useless. The most effective method for putting electric motors and generators out of operation is to injure the field or armature windings. This may be accomplished by breaking the motor shell with a sledge and using a crowbar on the coils. It time is not available for this procedure, a small-arms bullet may be directed into the coils through the air vents in either end bell. However, care should be taken to keep personnel from the path of a possible ricochet. Switch panels, plugs, and fuse and circuit-breaker panels should be smashed.

64. Optical Instruments

Sights and all other optical instruments should be evacuated if possible; if not, they should be smashed.

INDEX

	Paragraph	Page
Ammunition trucks, troughs, tables, and cranes. *See* loading mechanism.		
Antifriction devices, elevating and traversing:		
Barbette carriage M4 (16-inch gun)	37	89
14-inch gun railway mount M1920	52	164
General	35	78
155-mm gun carriage G.P.F.	47	135
155-mm gun carriage M1	48	145
Arc, searchlight	54, 56	172, 176
Arc length control, searchlights	54, 56	172, 176
Atlantic elevator equipment	36	85
Ballista	4	2
Barbette carriage. *See* carriages.		
Base ring:		
Barbette carriage M4 (16-inch gun)	37	89
8-inch gun barbette carriage M1	41	104
8-inch gun railway mount M1A1	50	157
8-inch gun railway mount M1918	51	162
14-inch gun railway mount M1920	52	164
General	35	78
90-mm gun barbette carriage M3	45	129
6-inch gun barbette carriage M1	42	113
Bofors (ogival) breechblock	14	18
Bogie, 155-mm gun M1	48	145
Bore, cannon tube	6	5
Bottom carriage. *See* carriages.		
Brakes:		
155-mm gun G.P.F.	47	135
155-mm gun M1	48	145
Brakes and overload slip devices, elevating and traversing mechanisms:		
Barbette carriage M4 (16-inch gun)	37	89
General	35	78
155-mm gun M1 and M1A1	48	145
3-inch gun barbette carriages M1902 and M1903	44	123
Breech recess	6	5
Breechblocks:		
Classes:		
General	13	18
Sliding-wedge:		
General	15	31
90-mm gun M1	15	31

	Paragraph	Page
Breechblocks—Classes (contd.):		
Slotted-screw:		
Carrier-supported type	14	18
8-inch gun Mk. VI Mod. 3A2	14	18
General	14	18
155-mm gun G.P.F.	14	18
155-mm gun M1	14	18
6-inch gun M1903A2	14	18
6-inch gun M1905A2	14	18
16-inch gun Mk. II M1	14	18
Stockett breechblock	14	18
Threading, methods of	14	18
Tray-supported type	14	18
Welin or step-threaded type	14	18
General	4, 11	2, 16
Obturation	12	16
Pressure gages	16	38
Built-up process, manufacture of cannon	4, 8	2, 7
Caliber, definition	5	4
Car body, 8-inch gun railway mount M1A1	50	157
Carriages:		
Fixed:		
Barbette carriage M4 (16-inch gun)	37	89
Barbette carriage M1919 (16-inch gun)	38	99
Barbette carriage M1920 (16-inch howitzer)	39	102
Barbette carriage M1917 (12-inch gun)	40	104
8-inch gun barbette carriage M1	41	104
General	33–36	74
90-mm gun barbette carriage M3	45	129
6-inch gun barbette carriage M1	42	113
6-inch gun barbette carriage, pedestal-type	43	121
3-inch gun barbette carriages M1902 and M1903	44	123
General	32	73
Motor-drawn:		
General	46	135
155-mm gun G.P.F.	47	135
155-mm gun M1	48	145
Railway artillery:		
8-inch gun railway mount M1A1	50	157
8-inch gun railway mount M1918	51	162
14-inch gun railway mount M1920	52	164
General	49	157
Carrier-supported type breechblocks	14	18
Casemate emplacement	34	74
Centering slope	6	5
Centrifugal-casting process, manufacture of cannon	8	7
Characteristics, seacoast artillery weapons	App. I	188
Cold-working process, manufacture of cannon	8	7
Construction of cannon tubes. See manufacture.		
Control stations, searchlights	56	176
Counterrecoil buffer:		
General	24, 28	55, 65
90-mm gun M1	31	69

	Paragraph	Page
Counterrecoil buffer (*contd.*):		
155-mm gun M1	28, 29	65
16-inch gun Mk. II M1	30	67
3-inch gun M1903	29	65
Counterrecoil or recuperator systems:		
General	24, 27	55, 63
Gravity	27	63
90-mm gun M1	31	69
Pneumatic	27, 29	63, 65
16-inch gun Mk. II M1	30	67
Spring	27, 29	63, 65
3-inch gun M1903	29	65
Counterweight of tubes	9	13
Cradle:		
Barbette carriage M4 (16-inch gun)	37	89
Barbette carriage M1919 (16-inch gun)	38	99
Barbette carriage M1920 (16-inch howitzer)	39	102
Barbette carriage M1917 (12-inch gun)	40	104
8-inch gun barbette carriage M1	41	104
General	9, 35	13, 78
155-mm gun G.P.F.	47	135
155-mm gun M1	48	145
6-inch gun barbette carriage M1	42	113
DeBange obturator	12	16
Demolition of materiel	57–64	184
Distant electric control searchlights	54, 56	172, 176
8-inch gun Mk. VI Mod. 3A2:		
Breechblock	14	18
Carriage M1, barbette	41	104
Carriage, railway mount M1A1	50	157
Firing lock Mk. VIII Mod. II	23	53
8-inch gun railway mount M1918	51	162
Electric firing of cannon:		
Firing lock Mk. I	18	38
Firing lock Mk. VIII Mod. II	23	53
Firing mechanism M1903	19	44
Elevating mechanisms:		
Atlantic elevator equipment	36	85
Barbette carriage M4 (16-inch gun)	37	89
Barbette carriage M1919 (16-inch gun)	38	99
Barbette carriage M1920 (16-inch howitzer)	39	102
Barbette carriage M1917 (12-inch gun)	40	104
8-inch gun barbette carriage M1	41	104
8-inch gun railway mount M1A1	50	157
8-inch gun railway mount M1918	51	162
14-inch gun railway mount M1920	52	164
General	35	78
90-mm gun barbette carriage M3	45	129
155-mm gun carriage G.P.F.	47	135
155-mm gun carriage M1	48	145
6-inch gun barbette carriage M1	42	113

	Paragraph	Page
Elevating mechanisms (*contd.*):		
6-inch gun barbette carriage, pedestal-type	43	121
3-inch gun barbette carriages M1902 and M1903	44	123
Emplacements, cannon:		
Casemate	34	74
8-inch gun railway mount M1A1	50	157
8-inch gun railway mount M1918	51	162
Firing platform, 155-mm gun M1	48	145
14-inch gun railway mount M1920	52	164
Open	34	74
Panama mount, 155-mm gun G.P.F.	47	135
Turret	34	74
Equilibrators:		
General	9	13
90-mm gun carriage M3	45	129
155-mm gun carriage M1	48	145
Extractors, 90-mm gun M1	15	31
Firing lock Mk. I	18	38
Firing mechanisms:		
Continuous-pull type	20	47
Firing lock Mk. I	18	38
Firing lock Mk. VIII Mod. II	23	53
Firing mechanism M1930	19	44
General	17	38
Inertia-type (90-mm gun)	21	49
155-mm firing mechanism (M1 and G.P.F.)	22	50
Fixed carriages. *See* carriages.		
Forcing cone	6	5
14-inch gun railway mount M1920	52	164
Gas check seat	6	5
Gas-ejection system, 6-inch guns, M1903A2 and M1905A2	42	113
General Electric searchlight. *See* searchlights.		
Gravity-type counterrecoil	27	63
Grooves, recoil brake	25	56
Grooves, rifling	7	6
Gun (rifle)	5	4
Guns:		
Breechblocks	11–16	16
Firing mechanisms	17–23	38
Tubes	4–10	2
History:		
Emplacements	34	74
Cannon	4	2
Recoil mechanisms	24	55
Trunnions (cannon support)	9	13
Howitzer	5	4
Hydropneumatic recoil system:		
General	29	65
155-mm gun M1	29	65

	Paragraph	Page
Hydrospring recoil system:		
General	29	65
3-inch gun M1903	29	65
Intensifier, 8-inch gun barbette carriage M1	41	104
Interior of cannon, subdivisions	6	5
Interrupted-screw breechblock. *See* b r e e c h b l o c k s, slotted-screw.		
Jacks, 8-inch gun railway mount M1A1	50	157
Lands, rifling	7	6
Limbers—155-mm gun G.P.F.	47	135
Loading mechanisms:		
Barbette carriage M4 (16-inch gun)	37	89
Barbette carriage M1919 (16-inch gun)	38	99
Barbette carriage M1920 (16-inch howitzer)	38	99
Barbette carriage M1917 (12-inch gun)	40	104
8-inch gun barbette carriage M1	41	104
14-inch gun railway mount M1920	52	164
Power rammer, general	35	78
6-inch gun barbette carriage M1	42	113
Manufacture of cannon tubes:		
General	8	7
Processes:		
Built-up	8	7
Centrifugal-casting	8	7
Cold-working (autofrettage)	8	7
Wire-wrapping	8	7
Mission, seacoast artillery	2	1
Mortars	5	4
Motor-drawn carriages. *See* carriages.		
Negative carbon feed, searchlights	54, 56	172, 176
90-mm gun M1:		
Breechblock	15	31
Carriage	45	129
Extractors	15	31
Firing mechanism	21	49
Recoil and counterrecoil systems	31	69
Obturation:		
DeBange system	12	16
Fixed ammunition	12	16
General	12	16
Ogival breechblock	14	18
155-mm gun G.P.F.:		
Breechblock	14	18
Carriage	47	135
Firing mechanism	22	50
Recoil and counterrecoil systems	47	135

	Paragraph	Page
155-mm gun M1:		
Breechblock	14	18
Carriage	48	145
Firing mechanism	22	50
Recoil and counterrecoil systems	26–29	59
Open emplacement, cannon	34	74
Outriggers:		
8-inch gun railway mount M1A1	50	157
14-inch gun railway mount M1920	52	164
Overload slip device. *See* brakes.		
Panama mount, 155-mm gun G.P.F.	47	135
Pedestal-type barbette carriage	35, 43, 44	78, 121, 123
Positive carbon rotation and feed mechanism, searchlights	54, 56	172, 176
Powder chamber	6	5
Power plants, searchlights	56	176
Power rammers. *See* rammers.		
Pressure gages	16	38
Racers:		
Barbette carriage M4 (16-inch gun)	37	89
8-inch gun barbette carriage M1	41	104
General	35	78
6-inch gun barbette carriage M1	42	113
Railway cars, ammunition and auxiliary	53	171
Railway guns and carriages. *See* carriages.		
Rammers, power:		
Barbette carriage M4 (16-inch gun)	37	89
Barbette carriage M1919 (16-inch gun)	38	99
Barbette carriage M1917 (12-inch gun)	40	104
General	35	78
Recoil and counterrecoil mechanisms:		
General	24–29, 32	55, 73
90-mm gun M1	31	69
155-mm gun M1	26–29	59
16-inch gun Mk. II M1	30	67
3-inch gun M1903	29	65
Variable recoil	26	59
Recoil and counterrecoil mechanisms. *See also* carriages.		
Remote control system M14 (6-inch gun)	36	85
Rifling	4, 7	2, 6
Rodman gun	4	2
Rotating bands, projectile	7	6
Safety features:		
Continuous-pull type firing mechanism	20	47
Firing lock Mk. I	18	38
Firing lock Mk. VIII Mod. II	23	53
Firing mechanism M1903	19	44
Firing mechanism, 155-mm guns	22	50
Searchlights:		
Fixed	55	176
General	54	172
Mobile	56	176

	Paragraph	Page
Shields:		
Barbette carriage M4 (16-inch gun)	37	89
90-mm gun carriage M3	45	129
6-inch gun barbette carriage M1	42	113
3-inch gun barbette carriages M1902 and M1903	44	123
Side frames. *See* carriages.		
6-inch gun M1900:		
Breechblock	14	18
Carriage	35, 43	78, 121
6-inch guns M1903A2 and M1905A2:		
Atlantic elevator equipment	36	85
Breechblocks	14	18
Carriages	42	113
16-inch gun Mk. II M1:		
Breechblock	14	18
Carriage	37	89
Firing lock Mk. I	18	38
Recoil and counterrecoil systems	30	67
16-inch guns M1919M2 and M3—Carriage	38	99
Slide rails, cannon support	9	13
Sliding-wedge breechblocks. *See* breechblocks.		
Slotted-screw breechblocks. *See* breechblocks.		
Spades. *See* trails and spades.		
Sperry searchlights. *See* searchlights.		
Splines, cannon support	9	13
Stress in cannon wall	4, 8	2, 7
Subcaliber guns and tubes	10	14
Tangential stress, cannon wall	4, 8	2, 7
Tapered breechblock	14	18
3-inch guns M1902M1 and M1903:		
Carriages	44	123
Firing mechanism	20	47
Recoil and counterrecoil systems	29, 44	65, 123
Throttling bars, recoil brake	25	56
Throttling grooves, recoil brake	25	56
Throttling rod, recoil brake	25	56
Tires. *See* wheels and tires.		
Top carriage. *See* carriages.		
Trails and spades:		
155-mm gun G.P.F.	47	135
155-mm gun M1	48	145
Transportation for mobile searchlights	56	176
Traveling position:		
155-mm gun G.P.F.	47	135
155-mm gun M1	48	145
Traversing mechanisms:		
Barbette carriage M4 (16-inch gun)	37	89
Barbette carriage M1917 (12-inch gun)	40	104
8-inch gun barbette carriage M1	41	104
8-inch gun railway mount M1A1	50	157
8-inch gun railway mount M1918	51	162
14-inch gun railway mount M1920	52	164

	Paragraph	Page
Traversing mechanisms (*contd.*):		
General	35	78
Hydraulic speed gear	36	85
90-mm gun barbette carriage M3	45	129
155-mm gun carriage G.P.F.	47	135
155-mm gun carriage M1	48	145
6-inch gun barbette carriage M1	42	113
3-inch gun barbette carriages M1902 and M1903	44	123
Tray-supported breechblocks. *See* breechblocks.		
Trunnions, cannon support	9	13
Turret emplacement	34	74
12-inch gun M1895:		
Breechblock	14	18
Carriage M1917	40	104
Firing mechanism	19	44
Twist of rifling	7	6
Types of cannon:		
Gun (rifle)	5	4
Howitzer	5	4
Mortar	5	4
Variable recoil	26, 32	59, 73
Waterbury hydraulic speed gear	36	85
Welin or step-threaded breechblocks	14	18
Wheels and tires:		
155-mm gun G.P.F.	47	135
155-mm gun M1	48	145
Wire-wrapping process, manufacture of cannon	8	7

31.3 Fort Monroe—10-15-44—3,350

APPENDIX I (page 1)
CHARACTERISTICS OF SEACOAST WEAPONS

GUN					
Caliber	Model	Length (Calibers)	Weight (Pounds)	Type	Rifling
3"	1902M1	50	1,950	Built-up	1 in 50 to 1 in 25
	M1903	55	2,690	Built-up	
90-mm	M1	50	1,465	Monotube	1 in 32
6"	M1900	50	19,114	Built-up	1 in 50 to 1 in 25
	M1903A2 M1905A2	50	20,700	Built-up	1 in 25
	M1908 M1908M1 M1908MII	45	12,300	Wire-wound	1 in 50 to 1 in 25
	MI	50	20,550	Built-up	1 in 25
155-mm	M1917A1 M1918M1	37	8,715	Built-up	1 in 29.9
	M1&M1A1	45	9,595	Built-up	1 in 25
8"	M1888 MII&MIII	32	33,200	Built-up	1 in 5 to 1 in 25
	MK. VI M3A2	45	42,00	Built-up	1 in 25
12"	M1895M1A4	35	114,700	Built-up	1 in 25
14"	M1920M1 M1920MII	50	192,500 233,800	Wire-wound	1 in 32
16"	M1919 MII, MIII	50	385,600	Wire-wound	1 in 30
	MK.II M1	50	307,185	Built-up	1 in 32
16"	M1920 (Howitzer)	25	195,300	Built-up	1 in 25

APPENDIX I (page 2)
CHARACTERISTICS OF SEACOAST WEAPONS

GUN				
Caliber	Model	M.V. (Ft. per sec.)	Life Full Charge (Rounds)	Maximum Range (Yards)
3"	1902M1	2,800	2,500	11,000
	M1903	2,800	2,500	11,300
90-mm	M1	2,700	900	19,500
6"	M1900	2,600	1,250	17,000
	M1903A2 M1905A2	2,800	1,000	27,100
	M1908 M1908M1 M1908MII	2,600	1,000	17,000
	MI	2,800	1,000	27,500
155-mm	M1917A1 M1918M1	2,410	3,000	19,100
	M1&M1A1	2,800	2,600	25,700
8"	M1888 MII&MIII	2,600 and 1,950	900	24,900
	MK. VI M3A2	2,840	300	35,300
12"	M1895M1A4	2,600	350	29,300
14"	M1920M1 M1920MII	3,000	200	48,200
16"	M1919 MII, MIII	2,700	175	49,100
	MK.II M1	2,750	175	45,150
16"	M1920 (Howitzer)	1,950	800	24,500

APPENDIX I (page 3)
CHARACTERISTICS OF SEACOAST WEAPONS

Gun		Breechblock		
Caliber	Model	Type	Thread	Number Handles
3"	1902M1	Carrier-supported	Plain	1
	M1903	Carrier-supported	Plain	1
90-mm	M1	Drop Block (Verticle Sliding Wedge)		Automatic-0 Manual-1
6"	M1900	Carrier-supported	Plain	1
	M1903A2 M1905A2	Carrier-supported	Plain	1
	M1908 M1908M1 M1908MII	Carrier-supported	Plain	1
	MI	Carrier-supported	Step	1
155-mm	M1917A1 M1918M1	Carrier-supported	Plain	1
	M1&M1A1	Carrier-supported	Step	1
8"	M1888 MII&MIII	Tray-supported	Step	2
	MK. VI M3A2	Tray-supported	Step	1
12"	M1895M1A4	Tray-supported	Plain	1
14"	M1920M1 M1920MII	Carrier-supported	Step	1
16"	M1919 MII, MIII	Carrier-supported	Step	1
	MK.II M1	Carrier-supported	Step	1
16"	M1920 (Howitzer)	Carrier-supported	Step	1

APPENDIX I (page 4)
CHARACTERISTICS OF SEACOAST WEAPONS

GUN		BREECHBLOCK	
Caliber	Model	Firing Mechanism	Method of Operation
3"	1902M1	Continuous-pull Percussion	Hand
	M1903	Continuous-pull Percussion	Hand
90-mm	M1	Percussion	Auto or Hand
6"	M1900	M1903 Elect. or Frict.	Hand
	M1903A2 M1905A2	M1903 Elect. or Frict.	Hand
	M1908 M1908M1 M1908MII	M1903 Elect. or Frict.	Hand
	MI	Elect. or Per.	Hand
155-mm	M1917A1 M1918M1	G.P.F. M1918 Percussion	Hand
	M1&M1A1	M1 Percusion	Hand
8"	M1888 MII&MIII	M1903 Elect. or Frict.	Hand
	MK. VI M3A2	Firing lock MK. VIII MII Elect. or Per.	Hand
12"	M1895M1A4	M1903 Elect. or Frict.	Hand
14"	M1920M1 M1920MII	Firing lock MK. I Elect. or Per. Combination	Air
16"	M1919 MII, MIII	Firing lock MK. I Elect. or Per. Combination	Air
	MK.II M1	Firing lock MK. I Elect. or Per. Combination	Air
16"	M1920 (Howitzer)	Firing lock MK. I Elect. or Per. Combination	Air

APPENDIX I (page 5)
CHARACTERISTICS OF SEACOAST WEAPONS

GUN	CARRIAGE			
Caliber	Model	Type	Weight (Pounds)	Elevation
3"	M1902	BC	4,075	-10° to +15°
	M1903	BC	3,310	-10° to +16°
90-mm	M3	BC	4,110	-8° to +80°
6"	M1900	BC	24,800	-5° to +20°
	M1	BC	75,300	-5° to +47.5°
	M1910	BC	30,000	-3° to +12°
	M3, M4	BC	67,450	-5° to +47°
155-mm	M1917M2 M1918M3	Mobile Split-trail	14,560	0° to +35°
	M1	Mobile Split-trail	20,305	2° to +63°
8"	M1918	Rwy.	140,800	0° to +42°
	M1	BC or	58,470	-5° to +45°
	M1A1	Rwy.	188,00	-5° to +45°
12"	M1917	BC	302,000	0° to +35°
14"	M1920	Rwy.	496,200	-7° to +50°
16"	M1919	BC	699,153	-7° to +65°
	M4	BC	665,315	-3° to +47°
16"	M1920	BC	705,000	-7° to +65°

APPENDIX I (page 6)
CHARACTERISTICS OF SEACOAST WEAPONS

GUN	CARRIAGE				
Caliber	Model	Traverse	RECOIL		
			Length (Inches)	Type Mechanism	Number Cylinders
3"	M1902	360°	9	Hydraulic	1
	M1903	360°	9	Hydraulic	1
90-mm	M3	360°	24 to 46	Hydraulic (Variable)	1
6"	M1900	360°	15	Hydraulic	1
	M1	360°	20	Hydraulic	1
	M1910	360° (Casemate 120°)	15	Hydraulic	1
	M3, M4	360°	18	Hydraulic	1
155-mm	M1917M2 M1918M3	60° (Panama Mt. 360°)	43 to 66	Hydraulic (Variable)	1
	M1	60° (Platform Mt 360°)	29 to 65	Hydraulic (Variable)	1
8"	M1918	360°	48	Hydraulic	1
	M1	360°	27	Hydraulic	2
	M1A1	360°(Casemate 145°)	27	Hydraulic	2
12"	M1917	360°(Casemate 145°)	30	Hydraulic	1
14"	M1920	Top Carr. 7° Car Body 360°	35	Hydraulic	4
16"	M1919	360°	40	Hydraulic	4
	M4	180°(Casemate 145°)	49	Hydraulic	1
16"	M1920	360°	38.5	Hydraulic	4

APPENDIX I (page 7)
CHARACTERISTICS OF SEACOAST WEAPONS

GUN Caliber	CARRIAGE Model	COUNTERRECOIL		
		Type Mechanism	Number Cylinders	Type Buffer
3"	M1902	Spring	1	Dashpot
	M1903	Spring	1	Dashpot
90-mm	M3	Hydropneumatic	2	Valve
6"	M1900	Spring	2	Dashpot
	M1	Spring	2	Dashpot
	M1910	Spring	2	Dashpot
	M3, M4	Spring	2	Dashpot
155-mm	M1917M2 M1918M3	Hydropneumatic	2	Dashpot and Valve
	M1	Hydropneumatic	2	Dashpot and Valve
8"	M1918	Spring	4	Dashpot
	M1	Pneumatic	1	Valve
	M1A1	Pneumatic	1	Valve
12"	M1917	Spring	4	Dashpot
14"	M1920	Pneumatic	1	Dashpot
16"	M1919	Pneumatic	2	Valve
	M4	Pneumatic	3	Dashpot and Valve
16"	M1920	Pneumatic	1	Dashpot

The Coast Defense Study Group

The **Coast Defense Study Group, Inc.** (CDSG) is a tax-exempt corporation dedicated to study of seacoast fortifications. CDSG's purpose is to promote and encourage the study of coastal defenses, primarily but not exclusively those of the United States of America. The study of coast defenses and fortifications includes their history, architecture, technology, strategic and tactical employment and evolution. The primary goals of the CDSG are the following:

* Educational study of coast defenses

* Technical research and documentation of coast defenses

* Preservation of coast defense sites, equipment and records for current and future generations

* Accurate coast defense site interpretations

* Assistance to groups interested in preservation and interpretation of coast defense sites

* Charitable activities which promote the goals of the CDSG

The CDSG was officially founded in 1985 at Fort Monroe, Virginia and incorporated in 1993. The first "St. Babs" conference was held in 1978 to visit the harbor defenses of New York. Subsequent annual CDSG conferences have been held at all the major harbor defenses of the continental Unites States. Several special overseas tours have also been organized. Membership is open to any person or organization interested in the study or history of the coast defenses and fortifications. Membership in the CDSG will allow you to attend the annual conference, special tours and receive the CDSG's quarterly journal and newsletter. For more information on the CDSG, please visit the CDSG website at cdsg.org or contact us at 24624 W. 96th Street, Lenexa, KS 66227-7285 USA, Attn: Quentin Schillare, Membership.

The **CDSG Fund** supports the efforts of the Coast Defense Study Group by raising funds for preservation and interpretation of American seacoast defenses. The CDSG Fund is seeking donations for projects supporting its goals. Donations are tax-deductible for federal tax purposes as the CDSG is a 501(c)(3) organization, and 100% of your gift will go to project grants. Major contributions are acknowledged annually. The Fund is always seeking proposals for the monetary support of preservation and interpretation projects at former coast defense sites and museums. A one-page proposal briefly describing the site, the organization doing the work, and the proposed work or outcome should be sent to the address below. Successful proposals are usually distinct projects rather than general requests for donations. Upon conclusion of a project a short report suitable for publication in the CDSG Newsletter is requested. The trustees shall review such requests and pass their recommendation onto the CDSG Board of Directors for approval. Send donations and grant requests to: CDSG Fund c/o Terry McGovern 1700 Oak Lane McLean, VA 22101-3326 USA or use your credit card via PayPal on the www.cdsg.org website.

This reprint of Army Technical Manual 4-210 was completed in 2018.

The CDSG ePress
The CDSG Digital Library

The CDSG has digitized an extensive set of historic manuals, reports, records and documents on the harbor defenses of the United States Army. The documents have been broken down to a general collection of manuals and reports (CDSG Documents collection) and also the records of the various harbor defenses so you can order sets of records by coast/harbor defense or get the complete collection (CDSG Harbor Defense Collection). The CDSG provides its back issues of the CDSG Publications in electronic format.

Back Issues CDSG Publications DVD costs $55 and annual update (with original DVD insert) $10.

The **CDSG Documents USB drive** covers a range of historical material related to seacoast defenses -- most are from the National Archives. Included are the annual reports of the chief of coast artillery and chief of engineers; several board proceedings and reports; army directories; text books; tables of organization and equipment; WWII command histories; drill, field, training manuals and regulations; ordnance department documents; ordnance tables and compilations; and the ordnance gun and carriage cards. This UBS drive of CDSG Documents costs $50. Order the CDSG ePress items directly from Mark Berhow at PO Box 6124, Peoria, IL 61601 USA or at berhowma@cdsg.org and through www.cdsg.org.

Documents related to specific harbor defenses. These PDF documents form the basis of the conference and special tour handouts that have been held at these locations. These collections are available as PDF files on DVD. They include RCBs/RCWs; maps; annexes to defense projects; CD engineer notebooks; quartermaster building records; and aerial photos taken by the signal corps 1920-40. Please consult www.cdsg.org for more details. Contact Mark Berhow by post/email if you are interested in ordering only specific titles.

The CDSG Press

Prices Include Domestic/International Postage $US currency only (cash, check, money order), allow 6-8 weeks for delivery - CDSG Books and CDSD Gear ($ domestic / $ International) or order by credit card from CDSG.org

Notes on Seacoast Fortification Construction by Col. Eben E. Winslow, 1920, 428 pp. 1994 reprint HC with drawings $45/$60

Seacoast Artillery Weapons Technical Manual (TM) 9-210 by U.S. War Dept. 1944, 202 pp. 2018 reprint PB $20/$30

The Service of Coast Artillery by F. Hines & F. Ward, 1910, 736 pp. 1997 reprint HC $40/$60

Permanent Fortifications & Sea-Coast Defences by U.S. Congress, 1862, 544 pp. 1998 reprint HC $30/$45

American Coast Artillery Matériel Ordnance Dept. Doc#2042 by U.S. War Dept., 1922, 528 pp., 2001 reprint HC $45/$65 *American Seacoast Defenses: A Reference Guide* (3rd Edition) by Mark A. Berhow, (2015) 732 pp. HC $45/$80

The Endicott-Taft Reports, reprint of original reports of 1886 and 1905 by U.S. Congress, 525 pp. 2007 reprint HC $45/$80

Artillerists and Engineers: The Beginnings of US Fortifications 1794-1815 by Col. Wade, U.S. Army. PB, 226 pp. $25/$40

CDSG Logo Hats each $20.00 domestic and $25.00 foreign.

CDSG Logo Patches each $ 4.00 domestic & foreign.

CDSG T-Shirts (XXXL, XXL, XL, L; Red, Khaki, Navy, Black) $18.00 Domestic and $26.00 Foreign.

Send Order to: CDSG Press Attn: Terry McGovern 1700 Oak Lane, McLean, VA 22101-3326 or order via cdsg.org

www.ingramcontent.com/pod-product-compliance
Lightning Source LLC
Chambersburg PA
CBHW031143160426
43193CB00008B/233